TERRORISM, RISK AND THE CITY

Terrorism, Risk and the City

The Making of a Contemporary Urban Landscape

JON COAFFEE
Global Urban Research Unit,
School of Architecture, Planning and Landscape,
University of Newcastle upon Tyne, UK

ASHGATE

Published by
Ashgate Publishing Limited
Gower House
Croft Road
Aldershot
Hants GU11 3HR
England

Ashgate Publishing Company
Suite 420
101 Cherry Street
Burlington, VT 05401-4405
USA

Ashgate website: http://www.ashgate.com

British Library Cataloguing in Publication Data
Coaffee, Jon
Terrorism, risk and the City : the making of a contemporary urban landscape
1.City planning - England - London 2.Terrorism - England - London - Prevention 3.City planning - Political aspects
4. Terrorism - Prevention
I.Title
307.1'216'09421

[illegible] of Congress Cataloging-in-Publication Data
Coa[illegible]n.
Terro[illegible] risk and the city : the making of a contemporary urban landscape / Jon Coaffee.
p. c[illegible]
Includes [illegible]ographical references and index.
ISBN 0-7[illegible]555-4
1. City planning--England--London. 2. Terrorism--England--London--Prevention. 3. Public buildings--Security measures--England--London. 4. Crime prevention and architectural design--England--London. 5. Risk perception--Engl[illegible]don. 6. Emergency management--E[illegible]

HT169.G72L353 2003
307.1'216'09421--dc21 2003056291

ISBN 0 7546 3555 4

Reprinted 2004

Printed and bound in Great Britain by Athenaeum Press Ltd., Gateshead

List of Tables

Preface

The late 1960s and early 1970s were a time of great change in both the way contemporary Western Cities were physically structured and managed. In particular, in America, there were dramatic developments in research looking at the relationship between urban design and violent and criminal behaviour, most notably through the publication by Oscar Newman in 1972 of *Defensible Space – Crime Prevention through Environmental Design*.

At this time, whereas the occurrence and fear of crime were causing major upheavals in thinking about American cities, in the UK it took more than a decade for researchers to begin to look seriously at similar parallels between design and crime. In the UK, there was a more pressing concern that had been periodically dominating political circles, namely the 'troubles' in Northern Ireland, and in particular the terrorist bombing of town and city centres by the Provisional Irish Republican Army (Provisional IRA). As such it did not take long for defensible space principles to be adopted and radically applied to urban areas in Northern Ireland such as Belfast and Londonderry in attempts to deter terrorism. This occured most notably through the setting up of so called 'rings of steel' – a series of tall steel gates which restricted access to the town centre. Subsequently, as the risk of terrorism began to subside, such approaches were scaled down.

Twenty years later, in the early 1990s, the development of defensive strategies encompassing the fortification and privatisation of the city were attracting significant attention. In the UK the 'Northern Ireland question' was once again dominating the headlines as the Provisional IRA set about a bombing campaign in England, most notably against Britain's financial core, the City of London. Following such attacks, approaches first adopted in Belfast were enacted in London in an attempt to prevent further bombings.

This book is concerned with chronicling the mechanisms and processes by which such attempts to 'design out terrorism' were instigated in London. In particular, it examines how the perceived risk of terrorist attack led to changes in the physical form and institutional infrastructure of the City of London between 1992 and 2003 during which the City was a prime terrorist target. This book explores the formal and informal strategies adopted by a number of key urban managers as they attempted to reduce both the physical and financial risk of terrorism through a series of place-specific security initiatives and insurance policies. It is shown that in central London the terrorist threat over the last decade has led to increased fortification, a substantial rise in terrorism insurance premiums and, changing institutional relations at a variety of spatial scales. It is argued that these changes were necessary to protect this area from further attack and to preserve the City's reputation as a global financial centre as well as London's position as a so-called world city. Furthermore, it will be argued that for both political and legal necessity the security measures deployed were advanced not in terms of an anti-terrorist effort, but in relation to the unintended by-products of such approaches, namely decreases in general crime, reduced levels of pollution and enhanced traffic management capabilities.

Research for this book started in the early months of 1995 during a period of relative calm in terms of Provisional IRA activity against English targets and has continued in and out of a number of 'ceasefire' periods. Whilst in 1995 the risk of terrorist attack by terrorists other than the Provisional IRA against the City of London was considered slim, the events of 9/11 have irreversibly changed this situation, leading to a scenario which can be described as one of 'postmodern terrorism' (Laqueur, 1996). This refers to a state of alert where the occurrence or threat of 'superviolence' through weapons of mass destruction (WMD) can have the capacity to terrorise and disable the functioning of economies and countries and radically alter the design of urban areas. Since 9/11, the City of London has once again been thrust into the forefront of London's and the UK's anti-terrorist response. This has necessitated a re-examination and updating of previous anti-terrorist policies and procedures, and, a commitment to develop new responses to deal with the ever changing terrorist threat whilst maintaining a balance between adequate security and business continuity.

Jon Coaffee
Newcastle upon Tyne, 2003

Acknowledgements

This book has taken a long time to complete and a large number of people have helped me along the way. In particular, I wish to thank John Gold and George Revill from Oxford Brookes University, and many colleagues at Newcastle University, for suggesting lines of reasoning and theoretical development during the course of this work. I am also indebted to many support staff within the School of Social Science and Law at Oxford Brookes, in particular Gerry Black for designing some of the illustrations for the book.

I also wish to thank all those from the fields of insurance, risk management, policing, politics and planning for their time and freely given opinions during the course of the research for this book, especially given the potentially sensitive nature of the topics being discussed.

A big debt of thanks must also go to Graham Soult for formatting the text for this book and to my family and friends for their constant encouragement.

About the Author

Dr Jon Coaffee is a Lecturer in Urban Regeneration within the Global Urban Research Unit, School of Architecture, Planning and Landscape, University of Newcastle upon Tyne. His main research interests revolve around contemporary urbanism and, in particular, the management of urban areas at different spatial scales.

He began working on issues related to the impact of urban terrorism in April 1995 as part of a doctoral research programme at Oxford Brookes University. This work continued when he moved to the University of Newcastle in 1998. He was awarded his doctorate in 2000 for work entitled *Risk, Insurance and the Contemporary Urban Landscape.* Recent global events, most notably 9/11, have necessitated a re-evaluation of this work, many ideas of which are presented in this book.

PART I
TRANSFORMING CITIES IN THE AGE OF TERRORISM

Chapter 1

Introduction: Terrorism, Risk and the City

It has recently been argued by many commentators that since the 9/11 attacks on New York and Washington the fortification and militarisation of the city has proceeded at an unparalleled pace (Davis, 2001; Light, 2002; Swanstrom, 2002). This trend is set to continue as, 'military doctrine and strategy will become more and more closely geared to the tactical and strategic protection of the political and economic key sites, zones and spaces of the global capitalist systems' (Graham, 2001, p.415).

However, it is important to note that the rise of military technology and strategy has always played a key role in the urbanisation process. Since the beginning of urban civilisation, defence against people or the natural elements has always been a factor influencing the structure and landscape of cities, becoming an ever-present preoccupation as the ruling powers sought to defend and secure their interests and create increased feelings of safety (Forbes, 1965; Postgate, 1992). As urbanisation proceeded so the defensive systems deployed by city authorities became increasingly sophisticated to cope with the improving strategies of intruders, in particular, through the construction of physical barriers in the form of gates, walls and ditches that were most widely used (Mumford, 1961; Dillon, 1994; Morris, 1994; Jordan *et al*, 1997). Such defensive structures, especially the city wall, also became associated with class distinction and the dual processes of inclusion and exclusion as the social élite lived within the defended citadel whilst the poor often lived in relative danger outside the city wall (Sjoberg, 1960; Pile *et al*, 1999).

In time the city wall became less important as a symbol of wealth, privilege and safety, as technological advances – most notably the invention of gunpowder – made such defences less effective (Keegan, 1993). Cities, however, continued to be characterised by defensive features as new walled and gated spaces developed, this time within the city boundaries, as danger was increasingly seen to originate from within, rather than outside, the urban area (Luymes, 1997; Atkins *et al*, 1998). As such by the mid nineteenth century many Western cities were characterised by secure residential estates amidst vast tracts of working class housing, which were seen as *terra incognita* (Newman, 1980; Jackson, 1992).

The Fortress City

Contemporary Western cities are no different from their predecessors in terms of making explicit attempts to use defence to structure the urban landscape. This situation is most pronounced in the United States where, since the 1960s, the relationship between defensive architecture and urban design has received widespread attention given rising crime rates and the declining state of high-rise residential dwellings (Jacobs, 1961; Boal, 1975; Gold, 1982; Newman, 1995). In particular, in the early 1970s the ideas of Crime Prevention through Environmental Design (Jeffery, 1971) and most notably Defensible Space (Newman, 1972) were popular approaches. Such approaches advocated 'designing out crime' through the addition or removal of physical features, which could control access, increase surveillance capabilities, and hence limit the opportunities for crime to occur in certain areas (Flaschsbart, 1969). Although the basis for this work was situated in an American context, during the 1970s and 1980s such approaches were extensively used by local authorities in the UK in existing housing schemes and in the design of new residential areas (Coleman, 1984, 1985; Dawson, 1984; Goodey and Gold, 1987).

In the 1990s further increases in violent crime, racial and cultural conflict, and material inequality within Western cities served increasingly to fragment the urban landscape creating 'radically new and complex logics of segregation and displacement' (McLaughlin and Muncie, 1999, p.117). This scenario was most commonly brought about through the superimposition of an array of fortification and surveillance devices on to the cityscape. Increasingly, residential areas, commercial centres, retail spaces, entertainment districts, and public facilities were fortified and privatised as the result of the actions of urban authorities, private businesses and wealthier citizens (Christopherson, 1994; Dillon, 1994; Flusty, 1994; Fyfe, 1997; Oc and Tiesdell, 1997). As Davis (1995, p.356) argued from an American perspective, 'we do indeed live in "fortress cities" brutally divided into "fortified cells" of affluence and "places of terror"'. More recently Lianos and Douglas (2000) articulated this idea through the concepts of 'dangerization' and suspicion where, they note, there is increasingly 'a tendency to perceive and analyse the world through categories of menace' (p.267) highlighting how the urban experience is seen through the differentiation of safe and unsafe areas:

> We are caught in a dynamics that increasingly colonizes the life-world through safe, controlled spaces and defines all non-monitored territories, and those who are in them as dangerous (p.270).

Through the 1990s the portrayal of the 'fortress city' was seen as one of the key characteristics of urban life and was, in large part, based on the experiences of Los Angeles, which assumed a theoretical primacy within urban studies *per se*, with a particular overemphasis on the militarisation of the urban landscape (Soja, 1989; Dear and Flusty, 1998; Dear, 1999). Of particular note was the work of Mike Davis on what he termed 'Fortress LA' (Davis, 1990, 1992, 1995, 1998). Davis depicted a city in which 'defence of luxury has given birth to an arsenal of security systems and an obsession with the policing of social boundaries through architecture. This militarisation of city life is increasingly visible everywhere in the built environment of the 1990s' (Davis, 1995, p.355). Such tendencies, although not to such extremes, can now be found in most urban areas, as recent studies in Brazil (Calderia, 1996; de Souza, 1996; Schiffer, 2002), Australia and New Zealand (Doekson, 1997), South Africa

(Worden, 1994; Sutcliffe, 1996; Napier, 2000) and the UK (Widgery, 1991; Harvey, 1996; Fyfe, 1998) demonstrate.

An example within a British context can be shown by Graham's (1995) work in Newcastle, which highlighted how a flagship office park planned for the run-down West End of the city could only attract occupiers 'if security was seen as a priority and CCTV was given an important role'. He noted how the proposed site was to be developed next to some of the most deprived neighbourhoods in Britain and that businesses, despite the promise of grant and rent subsidy, would not initially locate in this area. He noted that 'the solution was judged to be the "fortress approach" whereby the Business Park attracted clients on the basis of high security' (ibid. pp.10-11). Subsequently an array of fences, gates, and CCTV were added to the design of the proposed development. In addition the help of the local police force was sought to advise on where the placement of such fortifications would be most effective and how best use could be made of private security provision. As a result the business park, has suffered little crime and vandalism and has been able to obtain a large reduction in insurance premium, in contrast to the adjoining neighbourhoods, which are still plagued by high crime rates and dwindling property values.

Graham's work highlights a number of key trends relating to the management, fortification and surveillance of urban space that drove the defence of selected areas of the city forward in the 1990s. His work showed that both occupiers and property developers are now assessing building security at the design phase where the militarisation of commercial buildings and their borders become 'strongpoints of sale' (Flusty, 1994; Dear and Flusty, 1999). Graham's work also pointed to issues related to insurance reductions, the increasing influence of the police within the planning and building design process, the increasing role played by the private security industry and the juxtaposition within urban landscapes of controlled and regulated spaces and areas of disadvantage and poverty.

The trend towards increased urban fortification *per se* is related to enhanced perceptions of fear amongst urban dwellers (Blakely and Synder, 1999; Furedi, 2002; Swanstrom, 2002). Ellin (1997), for example, argued that 'form follows fear' in the contemporary city, with people in areas perceived to be at risk increasingly constructing defensive enclaves to protect themselves. In this sense the ability to pay for such fortifications is crucial. As Christopherson (1994, p.420) pointed out, 'there is no doubt that the new fortress-like environments respond to some version of consumer preferences'. As a result, many have argued that contemporary city life has been fundamentally reorganised as certain sections of society seal themselves away from the rest of the city creating new types of 'privatised' public space, which do not provide the same degree of access to all members of society (Sorkin, 1995; Lees, 1998; Gottdiener, 2000).

Within the Western city the desire for secure urban environments is now seen as one of the defining characteristics of so-called postmodern urbanism (Ellin, 1997; Dear, 1999) or the design-led approach of new urbanism (Harvey, 1997; Fainstein, 2000). Oscar Newman who coined the term 'defensible space' in the 1970s as a solution to urban crime can now be seen as one of the most influential thinkers on urban design issues in the United States (Harvey, 1996). Indeed, Newman himself writing in 1995 suggested that defensible space ideas could be rejuvenated and used as a new physical planning tool for urban revitalisation. There is also clear evidence that Newman's ideas continue to influence urban policy in the UK, which promotes neighbourhood renewal, social inclusion and a design-led urban renaissance. Increasingly, local authorities are now encouraged to form strategic partnerships with the police and local residents to reduce crime in their area with importance being placed on 'policies and guidance for designing out crime' (Urban Task Force, 1999, p.127).

Such an emphasis on countering crime through changes in urban design have for many years, in certain cities, also been linked to reducing the risk of terrorist attack. This concern has become ever more pertinent since 9/11 with many commentators drawing attention to the potential impact of 'new' terrorist threats in relation to the design and functioning of cities (Marcuse, 2000a; Marcuse and Kempton, 2002; Mills, 2002; Warren, 2002).

The Terrorist Threat

In recent years it has not just been the perceived risks of crime and intrusion that have led urban authorities, in collaboration with the police and the private security industry, to construct defensive urban landscapes. Increasingly, the potential threat of urban terrorism in certain cities has necessitated attempts to 'design out terrorism' through the addition of advanced security design features which have to be constantly updated to keep pace with the ever-changing terrorist threat (Haynes, 1995; Hyett, 1996; Hoffman, 1998). Previous studies on the impact of terrorism on the functioning of cities highlighted that if an urban area is vulnerable, or perceived by the community to be at risk from terrorist attack, then a reduction occurs in business confidence and public unease ensues (Compton *et al*, 1980; Brown, 1985b; Jarman, 1993). This, in the worst-case scenario, leads to business relocation from the threatened area or, the reluctance of the public to visit certain parts of the city. For example Brown (1984) showed how Belfast city centre was adversely affected by the Provisional Irish Republican Army's (Provisional IRA's) bombing campaign of the 1970s and the early 1980s, and, how business establishments were in favour of high levels of security to ease public concerns over safety. This starkly illustrated that in certain contexts terrorism can serve to exacerbate the concern for safety, which is already a key concern for users of the city, and lead to increased fortification of the urban landscape (Rycus, 1991).

Since the 1960s, fuelled by the growth of the mass media, terrorism has been widespread and associated with attacks against military establishments, government buildings, VIPs, or concentrations of particular racial or cultural groups (Livingstone, 1994; see also Picard, 1994; Wilkinson, 1997). In the early 1990s the global tendency of such targeting shifted noticeably towards economic targets, with the aim of causing economic disruption, social unease and putting direct or indirect political pressure on the ruling powers (Hillier, 1994; Rogers, 1996). Such targeting commonly began to be concentrated against business districts, gas and electricity plants, telecommunication infrastructures and transport networks.

In the early 1990s important financial centres became prime targets of attack because of their vast array of new 'designer' office buildings, their increasingly cosmopolitan communities, the damaging effects of bombing on commercial activities and the significant media attention and publicity that could be obtained by the terrorists. Examples of such commercial targeting included bombs in the financial districts of New York and Bombay in 1993, and Tokyo, Madrid, Paris, Riyadh and Colombo in 1995. In a British context, the main terrorist threat during the early-mid 1990s came from the Provisional IRA, with their prime target being the City of London (also known as the 'Square Mile' or 'the City') due to its symbolic value as the traditional heart of British imperialism and its economic importance at the centre of the British and global financial system. This book is specifically concerned with the impact and reaction to two vehicle-bomb attacks the Provisional IRA carried out in the City in April 1992 and April 1993, the indirect effect on the City of two bombings in 1996 in the London Docklands and central Manchester, and the more recent worldwide attacks on or after 9/11.

During the late 1990s the threat of such economic terrorism received a good deal of attention from international leaders. For example, in June 1996 the US President Bill Clinton called on world leaders to work together to combat international terrorism. On a similar note the British Prime Minister of the time, John Major, cited the Provisional IRA bombings in London and Manchester in 1996 and the Tokyo subway poison gas attack in 1995 as examples of how terrorism affects security and freedom and stated that 'it is a problem from which no one can hide and on which we must all co-operate. This is the security challenge of the 21st century'.[1]

In short, the threat of terrorist attack over the last decade has served to affect materially and symbolically the contemporary urban landscape in areas perceived to be at risk. Urban terrorism has created security threats to which municipal and national governments were forced to respond in order to alleviate the fears of their citizens and business community. As a result security measures similar to those used to 'design out crime' have been increasingly introduced, including physical barriers to restrict access, advanced surveillance techniques in the form of security cameras, and insurance regulations and blast protection, as well as innumerable indirect measures that operate through activating individual and community responses.

However, as will be highlighted in this book, after 9/11 both the perception of what constituted terrorism and the subsequent counter-responses changed dramatically, as a result these acts of mega-terrorism in New York and Washington and future fears about so-called 'postmodern terrorism' and the use of weapons of mass destruction (WMD). As Saifer (2001, pp.42-3) noted:

> The attack on New York has been by far the most hideous and devastating of that class of terrorist outrages which have which have been conceived and implemented by dedicated terrorist groups and networks on a 'demonstration' basis, in circumstances where otherwise vigilant security systems have foiled many other attempts. Death and destruction have been visited on cities and their citizens in the same way, by terrorist attacks on individual buildings and urban spaces, in such places as... the central area of Manchester and the financial district of the City of London. The attack on the World Trade Center expanded the scale and the 'global reach' of such 'demonstrations' by a truly appalling margin.

As such new forms of terrorism have subsequently led to new levels or urban vulnerability, and hence new forms of protective security in attempts to counter the threat in many urban centres.

The Book Structure

This book proceeds in the light of recent debates about the risk and uncertainty of urban life and the spatial restructuring of contemporary cities relating to the control and organisation of city spaces. The focus on this enquiry is on a singular case study of the City of London and will highlight the strategies that were adopted by key agencies such as political authorities, financial institutions and security professionals to exert control over selected areas of the city for the purpose of reducing the perceived risk of terrorist attack The time period covered by this study spans over a decade dates from the time of the first major City bomb in April 1992 until February 2003 when the anti-terrorist security measures developed in the City were integrated into part of a wider central London system for congestion charging.

This book addresses a number of key issues. First, it highlights the need for, and the consequences of, the high-levels of security that were deployed in the City of London due to the risk of terrorism. In particular, it will show how the physical form of the urban landscape was changed due to counter terrorism initiatives and how these impacted upon the functioning of the Square Mile and its neighbouring areas. It will detail how these physical alterations evolved over time and the reasons why particular security designs and features were adopted. In this context, emphasis is placed on the historically and geographically contingent nature of the City. For example, the City in previous eras has sought to defend itself against intrusion and attempts to deter terrorism in the 1990s and early 2000s are seen as the latest example of this defensive tendency. This aspect of the book will also analyse the similarities and differences between the defensive approaches utilised in the City of London and other security-related approaches that are often highlighted as a signature of the postmodern city. In particular, it will be argued that the 'Fortress urbanism' model, so prevalent in America, is of only limited importance in the British context where subtler forms of defended landscapes are commonly adopted.

Second, the book highlights how these anti-terrorist security measures were developed and activated by a distinctive set of local governance arrangements and global processes mainly related to the economic functioning of the Square Mile. This will show how the tightly-knit institutional arrangements in the City were crucial in the decision to develop anti-terrorism security measures, and indicate how the power embodied in these networks served to exclude alternative views or criticism of the development of high levels of security in the City. In other words, this book highlights the contested ways in which the construction of defensive landscapes in the City was viewed by different groupings both within and outside the City. To do this it draws on versions of recent 'institutional theory' which has highlighted the importance of institutional arrangements in a variety of social settings (Giddens, 1984; Beck, 1992a; Amin and Thrift, 1994; Healey, 1998). Such work in particular has drawn attention to the tendency for powerful 'voices' to dominate the urban planning agenda and, in so doing, marginalise alternative visions of development (Raco, 1998; Macleod and Goodwin, 1999).

Third, this enquiry investigates the influence of the regulatory role played by the insurance industry in shaping the landscape in the City as a result of enhanced terrorist risk. Traditionally, most previous work in British urban studies has neglected the role of the insurance industry as a regulator of the urban landscape focusing instead on the industry's investment practices (Cadman and Catelaeno, 1983; Henneberry, 1983; Faulsh, 1994; Murray, 1994; Doornkamp, 1995). Recently the role of insurance has received considerable attention in recent academic accounts of the influence of environmental and manufactured risk on social and economic relations (Giddens, 1990, 1991, 1994; Beck, 1992, 1995, 1999, 2000; Adams, 1995).

During the 1990s it became increasingly evident that high impact risk events (such as terrorism, global warming, ozone depletion, and earthquakes) were beginning to worry the insurance industry. In some cases this forced them to withdraw from the specific markets concerned, citing the high cost of their liability as well as the impossibility of calculating the risk involved. It was not that the frequency of such events increased, rather it was the insured cost that rose exponentially leading to fears of insolvency among insurers. For example, Giles (1994) cited a report by the Chartered Insurance Institute, which warned that British insurers were only now beginning to realise the scale of risk, they were 'carrying on their shoulders'.[2] This is the scenario that faced the insurance industry after the first major terrorist bombing of the City of London in April 1992.[3]

This part of the book specifically focuses on the evolution of terrorism insurance in the UK. In particular, it assessed the role of various insurance associations and representative bodies in bringing about a suitable terrorism insurance scheme and how this sought to redistribute the financial risk of terrorism away from the City, providing financial security to businesses in the wake of terrorist threats. Furthermore the relationship between insurance and counter-security measures within the City is assessed to show how terrorism insurance policies influenced and reinforced the proliferation of physical security measures within the Square Mile.

Overall, this book is structured into three main parts. Following the introduction, the remainder of Part I (Chapters 2-4) considers a number of conceptual and contextual ideas.

Chapter 2, *Urban Restructuring and the Development of Defensive Landscapes*, initially notes the historic nature of defensive cities. It relates their features to the changes that have occurred within the structure and functioning of major cities in the past thirty years as a result of the dual trends of the militarisation and the privatisation of public space. It does this through the use of the concept of territoriality. This chapter also examines the response of the Belfast authorities in the 1970s to the threat of terrorism.

Chapter 3, *Controlling the Security Discourse in the Postmodern City*, depicts the way in which institutional networking and partnership are increasingly serving to structure urban governance agendas with a particular emphasis on the merging of economic competition and security agendas. This will draw on contemporary examples from North America (in particular Los Angeles) and Britain to show how the police, the business community, the local government as well as other key urban stakeholders are attempting to promote security in specific localities to reduce the fear and occurrence of crime. It is argued that this is achieved through a combination of the managerial and regulatory strategies undertaken by the agencies of security, fortification techniques and, enhanced surveillance capabilities.

Chapter 4, *Risk Society and the Global Terrorist Threat*, introduces contemporary risk theory through the recent work of German sociologist Ulrich Beck and others, showing that Western society has created a scenario where new and destructive forms of risk have now become a major concern and are often deemed commercially uninsurable. The chapter then discusses contemporary risk within a specifically urban context by examining how the insurance industry has responded to this growing threat in the city through attempts to redistribute financial risk, and the insistence upon advanced risk management as a precondition to granting policy cover. Finally, this chapter deals directly with the ideologies underlying contemporary terrorist risk and relates this specifically to the British and Irish context, both historically and during the 1990s, before finally discussing new forms of terrorist threat in the post 9/11 era.

Part II of this of this book contains Chapters 5-7, which outline and discuss the main findings of this enquiry. Chapter 5 explores the role of the agencies of security, most notably, the City of London Police, in constructing a security cordon and a number of other security initiatives, in and around the Square Mile. Chapter 6 notes the influence of the insurance industry as it attempted to protect the City from the financial risk of terrorism. This chapter will also consider the relationship between terrorism insurance and an increasingly fortified landscape. Chapter 7, by contrast, describes the critical role of the Corporation of London, (the Local Authority for the Square Mile) in facilitating the enhancement of physical security for the City. It shows how local pro-security strategies were seen as essential to allow the Square Mile to remain competitive in the global economy, and how the powerful influence of the Corporation of London dominated arguments about how to respond to the terrorist threat.

Part III of the book includes Chapter 8 and 9. Chapter 8, *Beating the Bombers: a Decade of Counter Terrorism in the City of London*, summarises the key findings in Part II and situates the experiences of the City within a wider theoretical and empirical frame. Chapter 9, *Terrorism and Future Urbanism*, by contrast, initially moves away from the experiences of the Square Mile and highlights a variety of possible urban scenarios that were depicted in the aftermath of 9/11. It then reintroduces the current anti-terrorist strategies in place in the City and argues that a balance has been successfully achieved and maintained between security and business functioning without having to resort to the overt fortressing strategies that many have recently advocated.

Notes

1 Cited in Jones (1996).
2 In short, insurability has become a critical issue especially within the reinsurance world. This affects the insurance industry, as by reinsuring part of their initial risk insurance companies have traditionally been able to underwrite large risks without fear of bankruptcy.
3 The removal of reinsurance meant that in November 1992 the Association of British Insurers advised its members not to underwrite terrorism insurance on the British mainland. Subsequently the Government were forced to become involved. This led to the formation of Pool Reinsurance (Pool Re) in December 1992. This was a government-backed company set up to underwrite the risk of terrorism. This provided financial security to businesses located in areas at threat from terrorism, most notably the City of London. This will be discussed in detail in Chapter 6.

Chapter 2

Urban Restructuring and the Development of Defensive Landscapes

Introduction

In the last thirty years the urban landscape has been increasingly restructured as a result of geographical processes at both local and global levels, which continue to divide the city into a series of independent territories, societies, cultures, and economies (Soja, 1989; Harvey, 1990; Davis, 1992; Graham and Marvin 2001). This transformation involves the dual processes of fragmentation and agglomeration, which can collectively be termed restructuring.

However, there is perhaps nothing new about fragmented and divided cities and attempts by certain sections of city society, especially the rich and powerful, to cluster in certain territories for the social and economic advantages it brings, such as security, economies of scale, and cultural solidarity (Soffer and Minghi, 1986; Atkins, 1993; Marcuse, 1993; Harvey, 1996). This has formed distinct *territorial enclaves*, within the urban landscape – the development of which has intensified in recent years, given the increasing growth of economic competition between cities, and as a result of the increased occurrence, and fear, of crime and terrorism in urban areas (Cheshire and Gordon, 1993; Fischer, 1993; Budd, 1998; Glassner, 2000; Furedi, 2002). As Jepson and MacGregor (1997, p.1) contended, 'relationships of contest, conflict and co-operation are realised in and through the social and spatial forms of contemporary urban life...[which] are creating new patterns of social division, and new forms of regulation and control'.

The Privatisation and Militarisation of the City

Privatisation and 'militarisation' are commonly highlighted as key features of contemporary urban life where 'form follows fear' (Ellin, 1997) often leading to changes in the physical form of landscapes as a result of increased perceptions of crime, terrorism or external attack, emanating from occupants in a particular area (Davis, 1992, 1998; Dillon, 1994; Archibald *et al*, 2002; Graham, 2002; Marcuse and Kempton, 2002). As a result, a whole plethora of fortified landscape features can now be found, or are planned, in many Western cities. Such features range from the simple removal of benches and other amenities to stop the homeless living on the street to the other extreme of gated and heavily guarded residential and commercial areas (Davis, 1990; Flusty, 1994; Jones and Lowrey, 1995; Dear and Flusty, 1998; Graham and Marvin, 2001).

The control of the urban fabric by the higher socio-economic groups has always been a characteristic of cities but in recent years, and especially after 9/11, this control has been increasingly asserted through an array of evermore sophisticated physical and technological measures which have the explicit aim of excluding the sections of society deemed a threat to a particular way of life. The popularity of security features expresses the privatisation of space according to the preferences of the rich and powerful, and subsequently fuels the growth of the private security industry (Sorkin, 1995; Zukin, 1995; Lees, 1998). As Flusty (1994, p.67) noted whilst talking about the erosion of what he termed 'spatial justice' in Los Angeles:

> Traditional public spaces are increasingly supplanted by such privately produced (although often publicly subsidised) "privately owned and administered spaces for public aggregation" such as shopping malls [and] corporate plazas...In these new post-public spaces, access is predicted upon real or apparent ability to pay.

Flusty continued by indicating that such changes in the urban landscape were inherently related to economic productivity – 'in such spaces, exclusivity is an inevitable by-product of the high levels of control necessary to ensure that irregularity, unpredictability and inefficiency do not interfere with the orderly flow of commerce' (ibid.).

The remainder of this chapter is divided into three parts, and will highlight how, and why, such defensive features are becoming increasingly prevalent within the landscape of many Western cities. The first part will describe the historical trends of defending certain areas of the city, noting that defence has always been a pre-occupation in urban areas. Second, this chapter will consider post-war attempts to design-out-crime in urban environments, which were in large part based on findings from research into human territoriality. This will be exemplified by a study of Belfast in the 1970s which increasingly sought to 'design out terrorism' through the construction of an array of access restrictions and surveillance measures. The third part of the chapter will explore the contemporary meaning of territoriality as applied to the city, indicating that it should be used as no more than an analogy for describing the fragmentation of the city for defensive purposes.

Urban Security

Cities have always been characterised by feelings of insecurity, invasion by competing groups within society, and the fear of crime. This has meant that the need for defence against external attack has been an ever-present pre-occupation (Chermayeff and Alexander, 1966; Morris, 1994). Archaeological records show that the early urban areas on the floodplains of great rivers such as the Nile, Tigris, Euphrates and Yangtze were often surrounded by walls, ditches and other defensive features to delimit the 'known' from the chaos and danger of the outside world (Postgate, 1992; Keegan, 1993). For example, Jericho was one of the first examples of a defensive city, which, in around 7000 BC, had a defensive wall and large bastion towers supported by a nine metre ditch which deprived attackers of a means by which they could approach the city walls (Atkins *et al*, 1998). Atkins *et al* also cite the Mesopotamian service town of Uruk, which in about 3000 BC built a new 9.5 km defensive wall to cope with increased urban populations and the risk of attack. A further example of an early defensive city is the Mesopotamian city of Ur which between in 2100-1900 was surrounded by a 26 feet high and 77 feet thick perimeter wall (Morris 1994). This wall essentially served to control access to the area from intruders. Both Uruk and Ur were essentially city-states and controlled territories.

These defensive features – a combination of a wall, tower and ditch – became the universal blueprint for the fortified city, a design that was to change little between the building of Jericho and the introduction of gunpowder some 8000 years later (Keegan, 1993). Keegan also pointed out that it would be wrong to assume that the defensive features of Jericho were widely used. He noted that the people of Jericho at this time were very rich and that only similarly wealthy city-states could afford to implement such fortifications.

What is clear is that where centralised modes of governance began to be established, construction of strategic defences around cities or regions became widespread. As Gold and Revill (1999, p.230) noted, these defences served to secure 'the *interests* of an imperial power, serving to establish a presence and create and image of power that might impress an indigenous population of rival colonialists'. The development of such urban assemblages inside city walls meant that the urbanisation process could therefore be read as an agent of social control (Wittfogel, 1957). In particular, the threat of external attack meant the ruling class could justify the dense concentration of population into an easily regulated space (Atkins *et al*, 1998).

As cities developed, defensive systems became more complex to cope with the improving strategies of intruders. For example Lanciani (1968) indicated that Ancient Rome was fortified seven times by seven different lines of walls between the fifth century BC and the third century AD with outer walls being set beyond inner ones. This process has been termed 'multivallation'. Poyner (1983) also cited the castles and the walled towns of Medieval Europe as historic examples of this trend. Here internal defences, as represented by the fortress which dominated the centre of the city, and external defence, in the form of a wall, were key features. Atkins *et al* (1998) indicated that the developments in military technology at this time meant that defensive technologies improved. Designs for stone-clad castles were imported into Europe from Arabia, which formed the centre of new settlements as urbanisation spread within the safe confines of the city-wall. Durham and Newcastle in north east England, and the Italian cities of Florence, Venice, Milan and Rome provide good historic examples of such defended settlements.

The city wall served a defensive purpose but also became 'a symbol of the sharp distinction between the city and the country, and stood as formidable reminders of class distinction' (Dillon, 1994, p.10 – see also Fumagalli, 1994). In contrast to today it was the poor that were found located outside the city walls with the noxious trades and were, in effect, the first suburbanites (Sjoberg, 1960).[1] In time, the city wall and castle became less important, but still remained as symbols of wealth, privilege and power.

In a similar way, today's cities have their own particular expressions of defence that are equivalent, yet distinctly different, from historical examples. In the last thirty years such defensive measures have been brought about due to rising crime rates, the escalation of social conflicts related to material inequality, intensifying racial and ethnic tensions, the heightened fear of crime, and of particular relevance to this book, increased attacks by terrorist groups against the commercial and political infrastructure of cities. This has subsequently led to an increasingly sophisticated array of fortification, surveillance and security management techniques being deployed by urban authorities and the agencies of security. These urban trends are often described in terms such as the 'fortress city', the 'walled city', 'gated communities' or modern day 'panopticons' which focus on reducing access, enhancing surveillance and increasing the number of security personnel on patrol. As such, they evoke the notion of human territoriality which is related to the spatial control of a given area by certain social groups.

Territoriality: A Concept of Significance?

Social scientists' concern with space has led them not just to study the characteristics of individual places but also the processes that territorially divide or appropriate portions of space for specific purposes (Demko and Wood, 1994). In the last forty years the concept of territoriality has been used in diverse research fields, and according to Gold and Revill (1999, p.235) it has become a 'translation term' which 'allows connections to be made between disparate strands of research with common terminology and consistent threads of analysis, without needing to make assumptions that the phenomena under investigation are the product of similar processes that apply regardless of cultural context'. For example since the 1960s human territoriality has been studied in a number of contexts, including geopolitical change (Giddens, 1985; Agnew, 1987; Agnew and Corbridge, 1995; Anderson 1996), the impacts of economic restructuring (Scott and Storper, 1986, Robertson, 1992a, 1992b; Swyngedouw, 1997; Brenner, 1999), religious and ethnic conflicts (Rowley, 1992; Shirlow, 1998), an interest in localism and the role of place attachment (Sibley, 1990; Atkins *et al*, 1998), the need for cultural solidarity (Boal, 1996; Anderson and Shuttleworth, 1998) and, in particular, crime prevention (Jacobs, 1961; Coleman, 1984; Newman, 1995). More recently it has been argued that the concept of territoriality can help explain the impact of defence as a key feature shaping the contemporary urban landscape (Flusty, 1994; Harvey, 1996; Herbert, 1997a, 1997b, Coaffee 2000). As Gold and Revill (1999, p.232) noted, territoriality is 'an area of enquiry that examines the rationale for the creation and maintenance of defended spaces in human affairs'.

The concept of territoriality was originally developed at the beginning of the twentieth century to describe patterns of animal behaviour, in particular related to birds (see for example Howard, 1920). As Gold (1980, p.79) noted, 'developed by the work of ethnologists, territoriality is the name given to the processes and mechanisms by which living organisms lay claim to, mark, and defend their territory against rivals'. Territoriality was seen as an innate imperative linked to the need for privacy, safety and security (Morris, 1967; Eser, 1971; Porteous, 1976). During the 1960s a renewed interest in ethological studies meant that attempts were made to highlight links between human and animal behaviour, seeking to conceptualise the spatial relationships between areas and boundaries (Ardrey, 1966; Hall, 1966; Flaschsbart, 1969). Consequently, the development of behavioural geography in the 1970s led to widespread discussion as to the strength of the relationship between animal and human behaviour and whether or not territoriality was instinctively or culturally produced. In relation to human action, Gold (1980, p.80) therefore at this time defined territoriality as:

> A broad term that describes the motivated cognitive and behavioural states that a person displays in relation to a physical environment over which he wishes to exercise proprietorial rights, and that he, or he with others, uses more or less exclusively.

It was concluded by most commentators that the concept of territoriality could be utilised to gain insights into human behaviour, but that comparisons with animals were minor and analogous (Edney, 1976; Malmberg, 1980). For example, Gold (1980, p.89) noted that the relationship between human and animal behaviour 'is at best an analogy...[although it] is a fresh perspective on the way man regards his immediate spatial surroundings'. However, he later noted that 'when applied *carefully* as an analogy, territoriality affords insight into human spatial behaviour and provides a framework by which geographers can profit' (Gold, 1982, p.45). This however, required an appreciation that territories are constructed on a variety of spatial scales and formed by widely divergent processes and practices, and that the analogy between animal and human behaviour 'has its limits' and can lead to the 'territorial illusion' if 'human attributes and behaviours are oversimplified in order to fit frameworks best reserved for animals' (Gold and Revill, 1999, p.233).

This territorial analogy has most frequently been used in urban research to describe segregation in terms of 'conflict interpretation', with social groupings reacting to a hostile environment by evoking territoriality or creating 'turfs' to preserve the character of a defended area and to instil a sense of cultural solidarity (Clay, 1973).

An example of early work in this field was Peter Collison's *Cuttleslowe Walls* (1963), which provided an account of territorial behaviour describing how the friction between people living in adjoining middle and working class areas in Oxford, England, led to the erection of two formidable barbed wire-topped walls in the 1930s. Collison also highlighted two other examples – Cardiff (1955) and Dartford, Kent (1958), but these, he noted, represented extreme cases of segregation. The construction of defensive walls, as in this case, was considered rare at this time.

In the late 1960s and early 1970s, another city which experienced such 'walling' on a much larger scale was Belfast, this time due to religious differences between social groups. Boal's study of the Shankhill/Falls interface in West Belfast in the late 1960s perhaps provides one of the best known examples of human territorial behaviour, where Protestant and Catholic communities were kept apart by a series of physical barriers (defensive walls, or peace lines) which acted as territorial markers in the urban landscape (Boal, 1969, 1971). Consequently the residential geography of the city became fragmented into a series of inclusive 'religious enclaves' (see also Boal, 1975; Boal and Murray, 1977; Boal and Douglas, 1982; Dawson, 1984).

It was not just physical barriers that were seen to exemplify territoriality, although, undoubtedly the majority of studies focused on this aspect. As such, the use of symbolic territorial markers was also been heavily researched. For example, Ley (1974) illustrated the local geography of the fear of crime in inner city Philadelphia, by showing that certain areas where local inhabitants knew drugs were peddled were actively avoided due to the active demarcation of gang territories by graffiti. The distribution of graffiti in this case was mainly concentrated in two areas – first the centre of the territory, and secondly at its boundaries where space was actively contested (see also, Ley and Cybriwsky, 1974). Milgram (1970) did similar work in New York, producing a 'fear map' indicating areas where New Yorkers felt increasingly threatened. Such a sense of fear is however seen as time-dependent; either short term – for example areas which are seen as safe by day and unsafe at night; or can function over a longer period of time where areas of cities or whole cities get stigmatised (Tranter and Parkes, 1979). Such research was mirrored by more contemporary work by Davis (1990) in Los Angeles, by Campbell (1992) on urban housing estates in Britain with serious crime problems, and by Jarman (1993) on the use of sectarian murals as symbolic barriers in Belfast.

However, arguably the most notable research undertaken in the last thirty years on human territoriality was related to attempts to reduce criminal behaviour by designing out crime on public housing estates (Jacobs, 1961; Jeffery, 1971; Newman, 1972; Freedman, 1975; Ronchek, 1981). The following two sections will explore this work and present a detailed example of how the principles of defensible space were applied outside the residential context, to help counter terrorist attack in central Belfast in the 1970s.

Defence through Urban Design

In the late 1960s and early 1970s, defensive architecture and urban design were increasingly used in American cities as a result of research which indicated a relationship between certain types of environmental design and reduced levels of violence (R. Gold, 1970). There were concerns that enhanced urban fortifications were socially and economically destructive (in terms of economic decline of the city centre and social polarisation) and that the provision of security was becoming increasingly privatised as individuals, having lost faith with the public authorities to provide a safe environment, increasingly sought to defend themselves. Robert Gold (1970, p.153) noted that there was nothing new in this trend, as 'historically when political institutions have failed to protect the public, individuals have taken steps to safeguard themselves, their families and their properties. The present trend is no different in this regard'. Consumer preferences were thus increasingly beginning to influence the proliferation of anti-crime features at this time:

> The urban environment is being fortified today, not primarily by public decisions, but mainly through a multiplicity of private choices and decisions individuals make in our

> decentred society. The private market is responding to growing demand for an increasing range of crime control devices and other means of safety. In some cases, safety has become a commodity that is explicitly sold or rented with real estate.
>
> (ibid.)

As a result of such concerns, American urban planners and designers looked for strategies to reduce the opportunity for urban crime. This was a direct response to the urban riots which swept many US cities in the late 1960s, as well as the perceived problems associated with the physical design of the modernist high rise blocks which were seen as breeding grounds for criminal activity (see for example Jacobs, 1961; Newman, 1972, 1973).

Initially, C.Ray Jeffery (1971) developed an approach called *Crime Prevention through Environmental Design* (CPTED), which suggested that the design and arrangement of buildings could create environments which would discourage normal patterns of social interaction and encourage criminal behaviours. His key idea was that the built environment could be designed or changed to facilitate social cohesion among residents in order to deter crime by making criminal acts harder to commit or get away with. In short, opportunities could be reduced leading to a decrease in crime and fear of crime. As Jeffery (1971, p.178) stated 'in order to change criminal behaviour we must change the environment (not rehabilitate the criminal)'. This work was followed by a host of studies on architectural and design determinism, which highlighted how certain urban structures, could affect behaviour (Kaplan and Kaplan, 1978; Mercer, 1975; Rappoport 1977).

However it was the publication of Oscar Newman's (1972) *Defensible Space – Crime Prevention through Urban Design* that stimulated the most intense debates on the relationship between crime and the built environment. Newman highlighted, like others before him (see for example Wirth, 1938; Jacobs, 1961), that anonymity in the city ran parallel to rises in crime rate, and drew attention to the increasing sense of anonymity and danger that city life entailed, noting in particular that residents in high rise blocks did not appear to know each other, making neighbourhood organisation of crime prevention difficult. In his studies Newman did not rule out the use of security fences or electronic surveillance technologies, but relying on these measures was seen as a last resort if more subtle design solutions were unsuccessful.

Newman's work on housing estates in New York and St Louis, led to the concept of *Defensible Space*, which he saw as a 'range of mechanisms – real and symbolic barriers…[and] improved opportunities for surveillance – that combine to bring the environment under the control of its residents' (Newman, 1972, p.3). Defensible space was seen as the physical expression of a social fabric that could defend itself and could arguably be achieved by the manipulation of architectural and design elements.

Newman's basic assumption was that most criminals behave with some rationality, selecting targets in relation to perceptions of high rewards coupled with a low risk of getting caught. Thus deterring crime was fundamentally about giving would-be intruders a strong sense that if they enter a certain space they are likely to be observed and that they would have difficulty escaping. Newman concluded that outside spaces become more defensible if they are clearly demarcated (for example by 'grated' fences and shrubbery) and were well lit. By contrast, the construction of solid fences and tall hedges for example would serve as a potential hiding place for the criminal and should be avoided.

Newman proposed that four interrelated design features could create a secure residential environment: first, *territoriality*, which could be achieved by the zoning of public space in and around residential areas to promote a greater sense of community; second, improved *natural surveillance*, which could be enhanced – for example by the realignment of windows; third, the *image* of the building structures could be altered to avoid the stigma of public housing; finally, *milieux* could be enhanced where the environmental surroundings of residential areas could be altered so that they merged with areas of the city considered safe such as institutional and commercial areas.[2]

Newman's ideas were inexorably linked to the late 1960s and early 1970s and reflected the 'growing interest of the architectural profession in the relation between environment and behaviour, with some influence from rather popularised anthropology and ideas of territoriality' (Poyner, 1983, p.8). Poyner further highlighted that 'defensible space' was considered attractive at this time, because the 'emphasis was on the use of the environment to promote residential control and therefore somehow return to a more human and less threatening environment' (ibid.). In short defensible space offered an alternative to the target-hardening measures that were being introduced to American housing at this time.

Newman's ideas subsequently became popular, mainly in the United States, as a concept underlying the design of new residential communities, although Dawson (1984) indicated that Newman's principles were used extensively by many local authorities in the UK in the late 1970s and early 1980s (Clarke, 1980; Clarke and Mayhew, 1980; Coleman, 1984, 1985; Goodey and Gold, 1987). For example Alice Coleman (1984, 1985) argued that the physical design of high rise housing estates in London and Oxford had a noticeable affect on the behaviour of its residents and that 'more humane conditions' could be achieved by revising housing layouts and estate access to give residents more control over their local environment. This could be achieved, for example, by modifying entrances to staircases and lifts and by fencing in areas. These ideas became popular in the UK as many British local authorities 'Colemanised' their worst estates with enthusiasm but failed to combine such design changes with much-needed housing management, and community and employment initiatives. As Goodey and Gold (1987, p.130) noted, 'Colemanisation is fast becoming as significant a heading for an array of territory-creating measures as Newman's defensible space a decade earlier'. Such work highlighted that criminals often considerer the physical characteristics of the built environment prior to committing crime in an area, highlighting the effect of the environment on behaviour (Brantingham and Brantingham, 1981, 1984).

Newman's work has been criticised for its poor statistical analysis containing 'unverifiable assumptions about causal relationships between physical design and crime' (Gold, 1982, p57), and its omission of the interplay between social and physical variables. In particular, its focus on environmental determinism was seen as too much of a generalisation (Bottoms, 1974; Mayhew, 1979; Poyner, 1983; Madanipour, 1996; Tijerino, 1998). However, the relationships between urban design and behaviour remained unclear although many studies pointed to possible correlations (Coleman, 1984, 1985). For example, Coleman blamed the pathology of high-rise public housing estates directly on design deficiencies (Gold, 1997) which were a product of the modernist aim of social engineering.

Newman also assumed that residents would be willing to defend a public space and intervene in the event of criminal activity. Although it has been argued that territoriality 'is useful in explaining why an individual is willing to summon help...and why this quality can be eased by the built environment' (Tijerino, 1998, p.327), it would be wrong to assume that active defence is a characteristic of human territoriality (Edney, 1976). Thus territoriality may be helpful in indicating that individuals often seek to defend space, but fails to explain the subjective perceptual mechanisms through which this is brought about (Tijerino, 1998).

Defending Belfast

In the British context, defensible space ideas were to have wider adaptations than the residential context. For example, Boal (1975) argued, in relation to the need for anti-terrorist security in Northern Ireland, that the ultimate level of security provision in a city was defensible space with its emphasis on territoriality, existing alongside physical barriers. The ways in which changes in urban design attempted to reduce the risk of terrorist attack in Northern Ireland are of direct relevance to this book. In particular, it will be suggested that the principles underlying attempts in Belfast to contain terrorism in the 1970s were utilised in a modified form by the agencies of security in the City of London in the 1990s.

Belfast in the 1970s could be seen as a laboratory for radical experiments on the fortification of urban space. A number of distinct defended territories were created along sectarian lines to give the occupants of a defined area, or individual buildings, enhanced security. Defensible space was the order of the day. As Jarman (1993, p.107) commented, since the early 1970s:

> The apparent permanence of the conflict and the lack of any solutions acceptable to all parties has meant that the ideological divisions have increasingly become a concrete part of the physical environment, creating an ever more militarised landscape.

During the first years of the 'Troubles' (1968-1970), the commercial core of the city was seen as a relatively neutral space within the segregated sectarian landscape and was relatively unaffected by terrorism. All this changed in July 1970 when a large bomb was detonated in the area without warning. In the following years the defensive landscape change that occurred in central Belfast could be seen as the model that other towns in Northern Ireland adapted to their own local circumstances of place (see for example McKane, 1975 and McQuillen, 1975). For example, on 15 July 1972 concrete barriers were put in place around the shopping district of Londonderry/Derry by the British Army to seal off the centre from the Bogside, the western area of the city from which the majority of Provisional IRA attacks were believed to have originated. The area was not completely sealed off but rather the number of entrances were limited making control and searching of vehicles easier. The local Chamber of Trade who had seen business premises destroyed and damaged in previous attacks welcomed the moves.

The counter-terrorist security apparatus around Belfast city centre was first initiated three days after those in Londonderry/Derry. On the 18 July 1972 new traffic restrictions were imposed, without warning, as barbed wire fences were thrown across the main streets creating a number of defensive segments with access controlled by the British Army. The city centre in effect became a 'besieged citadel' (Jarman, 1993, p.115). This initially led to fears that these measures would destroy the city centre in a way the Provisional IRA never could, by keeping the customers out. Brown (1985a) pointed out the usual forces of city centre decline such as population dispersal, out of town shopping centres and increased car ownership which also, in part, contributed to the decline of central Belfast's retail core. Boal (1995, p.89) further commented 'the bombing campaign of the Provisional IRA, which appeared to be a concerted attempt to cripple the city's commercial life, led to the destruction of some 300 retail outlets and resulted in the loss of almost one-quarter of the total retail space'. The drastic security measures were taken due to the unsuccessful attempts by the authorities to tackle the security problem and can be seen as a radical example of territoriality as encompassed in notions of defensible space.

The bombing campaign against the city centre peaked in intensity on the 21 July 1972, three days after the construction of the cordon, when the Provisional IRA detonated twenty two bombs in and around Belfast city centre within the space of seventy five minutes. This day became known by all communities as Bloody Friday. As the *Belfast Telegraph* commented on 22 July (p.1):

> The city has not experienced such a day of death and destruction since the German blitz of 1941 [however]...it was significant that all yesterday's explosions occurred outside the new restricted traffic zones.[3]

This indicated that the Army and police's defensive strategy had been successful as far as defending the central business district was concerned. However, fears that the Provisional IRA would continue to plant bombs just outside the security cordon led to more Royal Ulster Constabulary (RUC) police officers being deployed in such areas. This highlighted a possible disadvantage of such a defensive landscapes, namely that by overtly securing one area, the would-be intruder, criminal or terrorist will seek out less well defended targets.

By 1974 the barbed wire fences encircling the central area had been replaced by a series of tall steel gates (the ring of steel), and civilian search units were established.

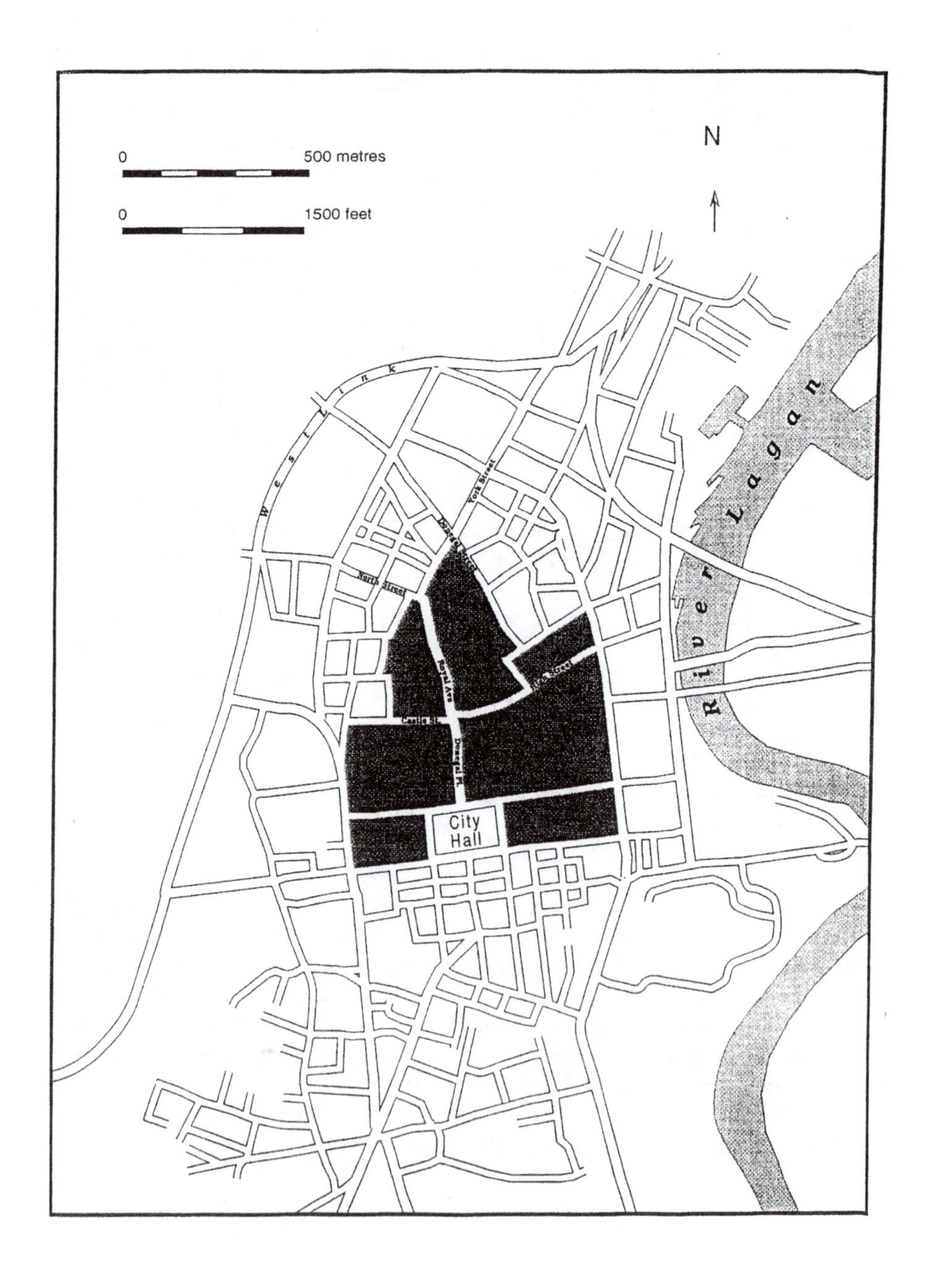

Figure 2.1 The initial security segments in Belfast city centre (1972/4)

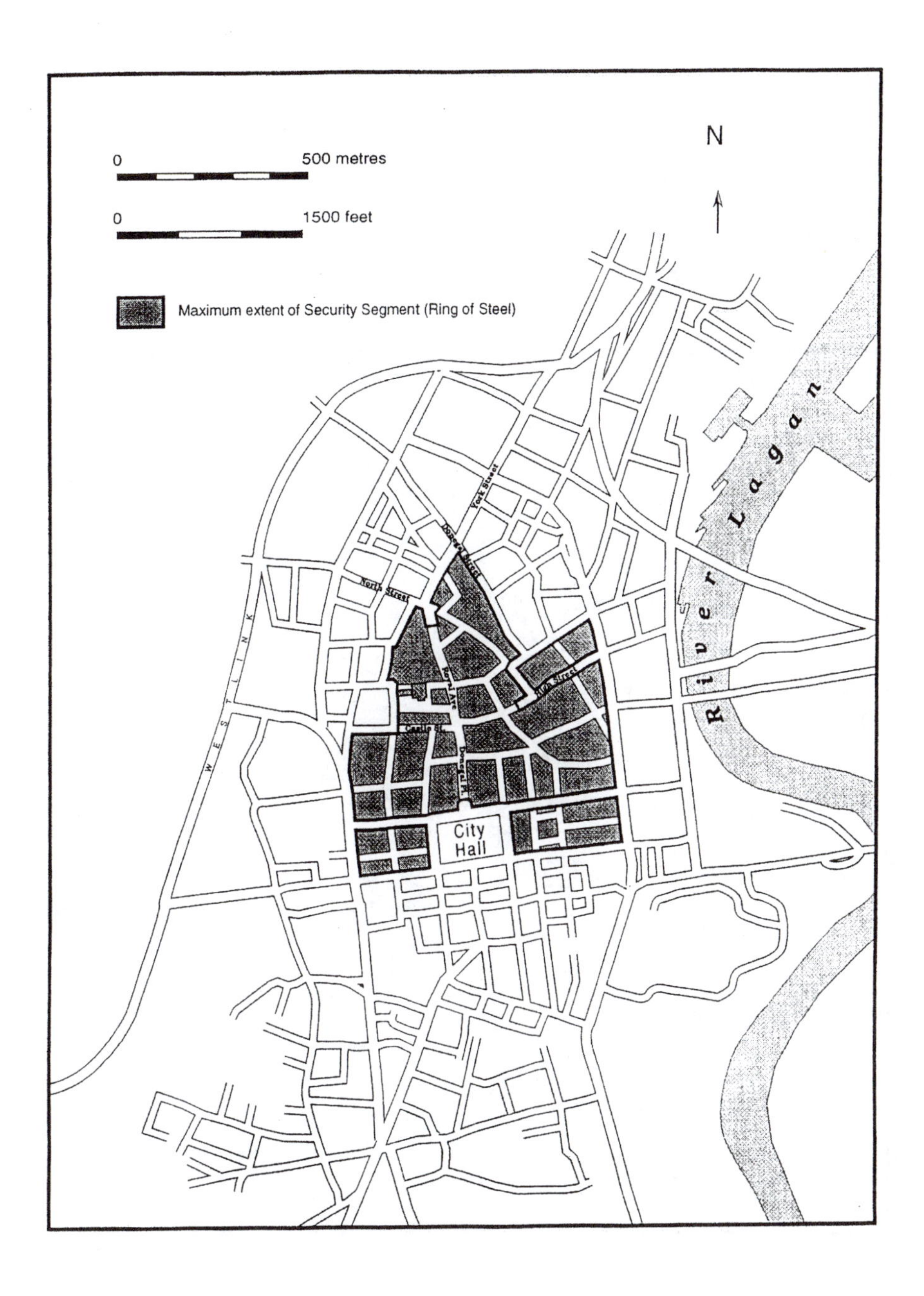

Figure 2.2 The maximum extent of the security segment (ring of steel) in 1976

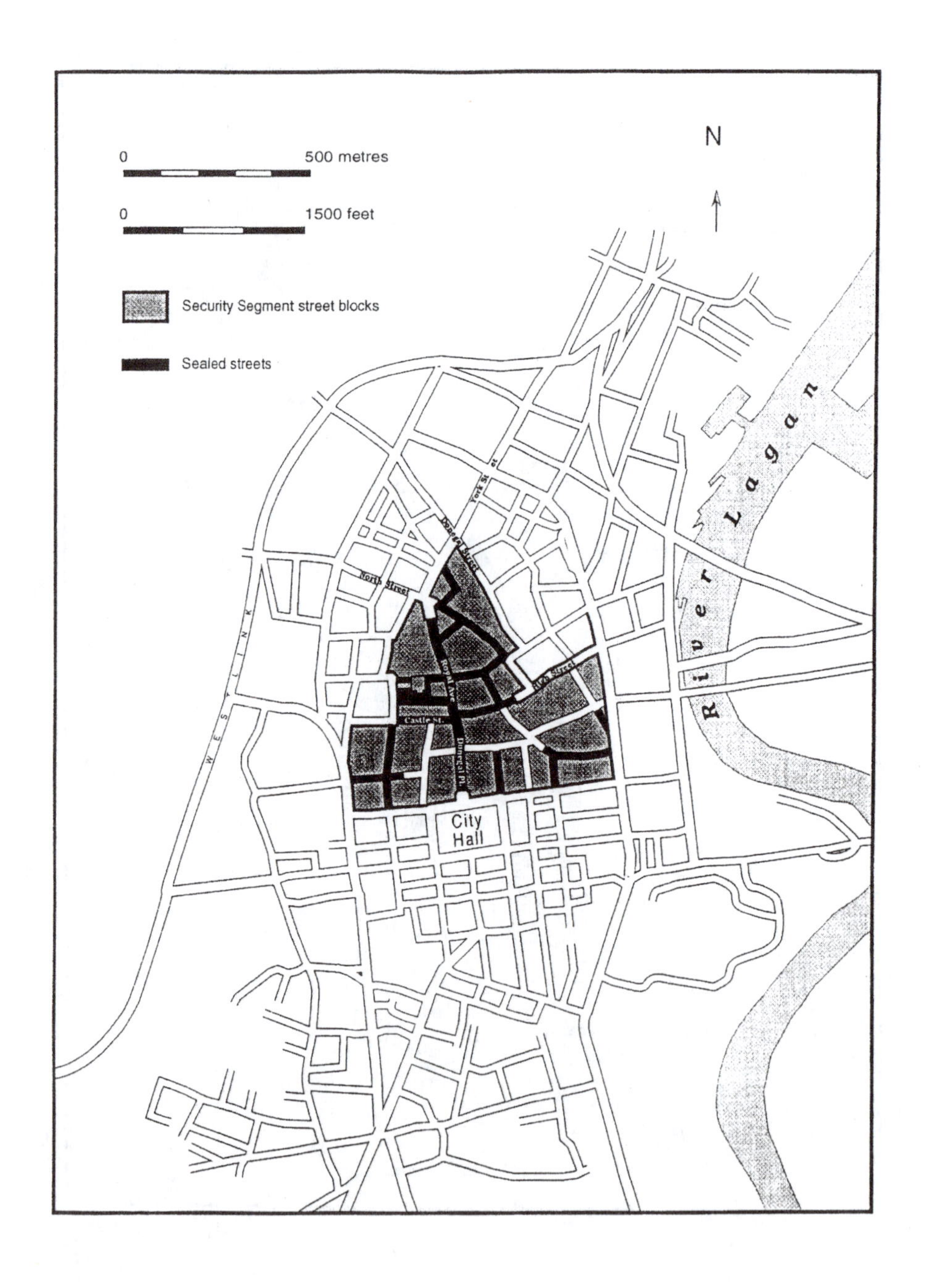

Figure 2.3 The Belfast ring of steel in the 1980s

Figure 2.1 shows the initial placement of the security cordon around Belfast city centre made up of seven different areas. Upon entering each area a bodily search was undertaken. Further advances were made in March 1976 when the four main security segments were amalgamated to form a single security zone (see Figure 2.2). This made shopping in the centre easier as only one search was required, although many shops employed their own searching teams to improve security. This arrangement also meant that the manpower needed to run the scheme effectively was minimised. There were only two vehicle entry points with the others being exit-only. It is believed that the term 'ring of steel' was first used in at this time to refer to the joining of the separate security zones into one large security sector ringed by seventeen 10-12 foot high, steel gates.[4] In addition, controlled parking zones were set up around the city centre in a further attempt to stop the car bombers outside the security cordon. For example, Compton *et al* (1980) described the formation of two types of parking zones, one covering the city centre where parking was strictly prohibited at all times (the so-called Pink Zone), and second, on the streets immediately surrounding the centre where cars could only be left in the evening and on Sunday (Yellow Zone).

Subsequently, as the risk of terrorist attack subsided, the three peripheral security segments to the south and north east were disbanded leaving just one main security zone (Figure 2.3).

Promoting the Defensive City

As the relative threat of terrorism against Belfast city centre decreased during the 1980s and 1990s, urban planners sought to re-image this 'pariah city' in an attempt to attract businesses back (Neill, 1992; Neill *et al*, 1995). Reduced levels of security, decreases in the number of terrorist attacks, redevelopment and pedestrianisation have subsequently helped to re-patronise central Belfast. Neill (1992) provided the following phasing of such planning ideas:

- *First*, the defensive policy of the seventies, which encompassed a radical defensive landscape to deter terrorist attack;
- *Second*, the encouragement of tentative development between 1980-1984, which saw a partial repatronisation of retail activity to the city centre; and
- *Third*, active promotion and planning from 1985-1992, which saw a concerted effort by the local authority to improve the physical infrastructure of the area as well as promote the area to business organisations world-wide.

However the re-imaging of Belfast has only been partially successful as it was often seen as superficial, and not as a catalyst to tackling the real problems of the city. Place promotion campaigns have also been used to re-image the stigmatised Central Belfast area. Gold (1994, p.23) refers to this type of advertising as the apotheosis of the traditional place marketing campaign to the point where the name Belfast was removed from adverts and was frequently replaced by Laganside or another development area name in an attempt to shake of its image of terrorist violence.

Jarman (1993, p.109) also commented on how Belfast is perceived by the outside world: 'The view from outside, largely mediated by television and newspaper photographers, is presented mostly as bomb-damaged buildings, endless parades and colourful paramilitary murals, and ignores their context within the more extensive transformations of the urban environment.' Neill (1992, p.9) concludes that Belfast city centre is a place where:

> A post-modernist consumerist kaleidoscope of images floats uncomfortably on top of the squat brutalism of terrorist-proof buildings and the symbolism of the past. It is a condition of visual schizophrenia.

In short the Belfast city centre came 'back from the dead' (Brown, 1985a, p.10). Brown qualifies this sentiment through a discussion, summarised in Table 2.1, which shows how Belfast city centre changed in the decade following the construction of the ring of steel. This indicates the direct positive correlation between a lack of terrorist bombs and business vitality.

Table 2.1 A decade of change in Belfast – 1974-84

	1974	**1984**
Bombings	62	3
Manned security gates	38	None
Off-street parking spaces	400	4500
Fully pedestrianised street	None	12
Private sector investment	Nil	£100 Million

Source: Brown (1985a, p.10).

Furthermore, Boal (1995, p.90-1) indicated that as terrorist violence continued to decline in the 1990s 'city centre accessibility was improved (new road works, multi-story car parking provision and some upgrading of public transport), while the environment of the centre was enhanced by extensive pedestrianisation and wide spread tree planting'. The New Year's bombing blitz of 1991-2 against the city centre and increases in security provision following major London bombings in the mid 1990s provided a testimony to the brittleness of Belfast's re-imaging.

The Success of the Security Cordon

The aim of the authorities and the security forces during the 1970s was to contain the terrorist threat, but there was also a consideration of the need to achieve a balance between security and the ability of businesses to trade. As Compton *et al* (1980, p.84) indicated, 'the traders tended to adopt an ambivalent attitude towards security measures, on the one hand wishing for increased security but on the other fearing loss of trade because of the curtailment of free access to the shopping area' (see also Hudson, 1973; Sheppard, 1976; Brown, 1985b). Since its inception, no car bomb has exploded inside the Belfast ring of steel and it can therefore, according to security forces, be judged successful. However, it was severely criticised by those traders just outside the cordon as they were increasingly attacked. For example, Compton *et al* (1980) highlighted the worry of the traders outside the cordon as well as those inside it and also stated that there were preferential trading locations within the cordon situated near the gates. The RUC indicated that as it attempted to balance security and commerce many establishments were 'choked-out' of business by the security cordon for which they received no compensation. Ironically it was often better for proprietors to be 'bombed-out' as compensation would be paid.[5]

Overall, Belfast provides an example of a city with a 'hardened' urban landscape where the spatial configurations within a defined area, especially borders and boundaries, have become more pronounced in an attempt to create territoriality. For example, this refers to ways in which buildings were delimited from their surroundings through defensive architecture, or ways in which groups have demarcated boundaries to separate their 'turf' from neighbouring ones. Furthermore, the security forces such as the police and army became 'major agents' in restructuring the urban landscape (Jarman, 1993).

However, landscape 'softening' also occurred in Belfast through the 1990s as attempts were made to remove territorial barriers and boundaries in order to increase access between previously conflicting territories given the reduced threat from terrorism, and Belfast's increased attempts to re-image itself within the global economy (Brown, 1987). Belfast's security cordon has, according to the RUC, forced the Provisional IRA to stop attacking the city centre on a regular basis. The Provisional IRA on the other hand would perhaps point to a change in overall tactics – away from attacking commercial property and towards attacking the security forces and the mainland (England) – as the reason why Belfast could begin to rebuild in the late 1970s and early 1980s.

In Belfast, the city centre became the focal point of both hardening and softening approaches. For example Jarman (1993, p.116) commented that whilst the British Government's approach to Belfast city centre was intimately linked to Provisional IRA tactics, it is also linked to what he called the Ulsterisation principle 'which principally involves criminalising the Republican movement while emphasising the uninterrupted continuation of daily life'. He continued:

> The city centre is the key to this approach, with the projection of an air of normality, accessibility and prosperity central to attempts to attract both British and foreign investment. A principal element in this strategy has been to remove the visible security presence...(which) has largely been reduced to security cameras.
>
> (ibid. p.116)

He concluded (p.117) by indicating that:

> The demilitarisation of the core emphasises the neutrality and impartiality of the commercial and administrative activities. The centre is projected as an area above and beyond the sectarian conflict.

The 'hardening' and subsequent 'softening' of the urban landscape of central Belfast over the past thirty years has been linked to an assumption that territoriality can be expressed though the built environment. This was initially done by the construction of walls, gates and security cameras, which served to exclude unwanted persons or activities from a defined area. These began to be slowly removed and it was not until mid-1995 that smaller security barriers replaced the remaining gates in the main shopping streets (see Figures 2.4, 2.5). Figure 2.4 show examples of the remaining security gates around Belfast city centre in 1995, whereas Figure 2.5 shows a new swing barrier-style gate, which replaced the old 1970s steel gates.

Indeed any decrease in security was offset by a centralised CCTV scheme, which became operational in December 1995. Cameras were seldom used at the height of the troubles due to expense and technological deficiency. The current system of cameras is similar to those adopted in many towns within Britain. Furthermore, the RUC during the Provisional IRAs cease-fire, from August 1994 until February 1996, indicated that they would have no compunction about resurrecting the ring of steel (and other security measures) if the cease-fire broke down and they considered there to be a realistic threat of the city centre being attacked.

During the 1980s, and 1990s the softening of the landscape as a result of the removal of much of this overt security apparatus, meant that a more symbolic approach to defending the area was undertaken as attempts were made to portray the city centre as a neutral space and encourage inward investment. Indeed during the late 1990s and early twenty first century Belfast began to actively market itself as a cultural centre of European significance. The aim of its leaders was to bid for the prestigious 'European Capital of Culture' crown which was to be given to a UK city in 2008. The 'failure' in 2002 to even get on the final shortlist of six cities for this honour perhaps reaffirmed that the 'troubles' have not gone away. For example, during the marketing campaign that led up to the short-listing the city had experienced continual bomb alerts, sectarian shootings and, within Northern Ireland as a whole, 'the peace process' was on the brink of collapse.

Towards New Definitions of Territoriality

During the 1980s social scientists began to use territoriality in a wider context moving away from the behavioural and ethological approach of previous decades. Perhaps the best-known work is Robert Sacks' *Human territoriality: its theory and history* (1986), which embraced a more symbolic and political definition of territoriality. Sack saw territoriality as 'socially and geographically rooted' and 'intimately related to how people use the land and how they organise themselves in space and how they give meaning to place'. Territorially in this sense can be seen as an expression of the interrelationship between space and society and was defined (p.19) as:

> The attempt by an individual or group to affect, influence or control people, phenomena, and relationships, by delimiting and asserting control over a geographical area.

Figure 2.4 Example of remaining security gates around Belfast city centre

Figure 2.5 A swing barrier at the entrance to Belfast city centre, which replaced the 1970s steel gates

In later work Sack (1992, p.52) argued that territoriality helped to embed meaning into place and organise spatial relations creating 'a personal sense of being in place to fixed locations in space' and showing that 'territoriality is a strategy for maintaining social order and imparting meaning to phenomena' (ibid. p.42). His definitions of territoriality had three interdependent spatial variables. First, a bounded spatial area that classifies the territory; second, the territory, often at its boundary, contains a form of communication identifying the territorial area; and third, direct attempts are also made to control access into the territorial area.

Sack argued that territorially is contingent on the particular history and geography of places and, importantly, that territoriality is a social want. Sack's ideas have however been criticised for overemphasising the general desire for, and apparently universal features of, human territoriality. Gold and Revill (1999, p.233) for example highlight the deficiencies and contradictions in Sacks' approach, noting that he 'falls foul of his own call for a historically and geographically sensitive approach...he works towards an abstract functionalist definition which itself becomes merely a set of empty and a historical categories when divorced from specific historical settings'.

Gold and Revill (1999) contended that territoriality be conceived as having 'its own cultural politics which, when deployed, serve the interests of certain groups rather than, and at the expense of, other'. This view, they noted, moves away from the idea of seeing territoriality primarily in terms of conflict resolution where space is contested, with the urban landscape being seen as being 'partitioned into mutually hostile units' (ibid. p.234). This moves towards a more inclusive definition where territoriality can be conceived in terms of physical expressions of conflict which overtly demarcate space, but can equally be viewed in terms of patterns of social and cultural organisation and rule making where individuals and groups can make 'important statements about the self and identity [which] are conveyed by the manipulation of controlled space and artefacts within that space' (ibid.). This under-emphasised aspect of territoriality studies is based on the notions of symbolic interaction (Cooper, 1974; Ericksen, 1980) where an area is seen as a 'landscape of social symbols' (Greenbie, 1982). In this case areas can perhaps be seen as symbolically displaying power and control over space to the rest of the city by what Taylor (1988) called 'territorial functioning'. This refers to how specific behaviours are a function of a sense of place which serves to enhance control of an area by clarifying spatial relationships in the area and by strengthening social bonding though a shared sense of identity and community (see also Ardrey, 1970; Gottdiener, 1995).

Conflict, Territoriality and the New Enclaving of the City

Whilst it is acknowledged that territoriality can be expressed through strategies of social and cultural organisation, it is most commonly articulated within the contemporary city in terms of defended enclaves of perhaps, two distinct kinds – *global* enclaves and *local* enclaves – which are commonly a result of the desire of higher-income residents and commercial enterprises to control and defend the space in which they live and work (Norfolk, 1994; Dillon, 1994, Urry 2001).

Global enclaves in this sense relate to processes linked to the world economy, which are causing agglomeration and territorialisation in financial areas of western cities. In particular, as a result of economic globalisation and the local processes of institutional change the contemporary urban landscape, put simply, looks different from its predecessor (Hubbard, 1996). In short, the landscape of large cities has become 'the terrain where a multiplicity of globalisation processes assume concrete, localised forms' (Sassen, 2000, p.147) or as Badcock (1996, p.94) argued, become filled with the 'designer spaces' of postmodernism. As Zukin (1988, p.435) noted, such landscapes 'directly mediate economic power by both conforming to and structuring norms of market-driven investment, production and consumption', This creates what can be referred to as 'landscapes of power', embodying such features as new-look buildings, high levels of place promotion and leading to greater asymmetry of power between a territory and its neighbouring areas. As Zukin (1992, p.197) concluded:

> The interrelated effect of economic structure, institutional intervention and cultural re-organisation are most directly perceived in change in the landscape: creating the city as a landscape of power.

The building cycle which enveloped the City of London in the mid-late 1980s and early 1990s, exemplifies this trend, and has important implications for this enquiry. The City's 1986 Local Plan stimulated a rush of planning applications to build new office complexes, and within eighteen months permission had been granted for twenty million square foot of office space to be built. This equated to a third of the total City's floorspace. Much of this construction was of new buildings and spaces of much larger size than the City had previously witnessed. The building of such 'groundscapers' was essential to fulfil the requirements of modern business, which required larger floorplates and height-ceiling ratios (Williams, 1991). It will be shown in subsequent chapters how the development of these new designer spaces, symbolic of the place of the City in the system of global finance, served to provide an incredibly attractive terrorist target, and also how the design of these structures and their surrounds were then subsequently altered to reduce the terrorist threat.

Despite the economic importance of new commercial buildings, there is however a noticeable downside to the urban restructuring efforts of local authorities in many cities in that 'the urban landscape is remodelled into visual spectacles of revitalised urban space and imagined community that mask real geographies of decay and neglect' (Goss 1997, p.181). As globalisation increasingly influences western cities, then urban complexity, as well as asymmetry, increase along class, ethnicity and gender lines. Injustice and inequality are frequently associated with the contemporary city, which is commonly seen as 'carceral' (Soja, 1989, 1997; Davis, 1990; Marcuse and Kempton, 2002).

Where there is power and wealth often there is found powerlessness and poverty nearby. Asymmetrical power has become an ever-present feature of city life, increasingly becoming a symptom within today's cities where the inequalities of capitalism are reproduced in the landscape. This can be amply illustrated in and around the City of London. For example, a report by Paul Valler (1999) in *The*

Independent illustrates this point by equating the cheek-by-jowl proximity of wealth and poverty at the borders of the City of London with areas of the Third World. Furthermore, Goodwin (1995) cited the differences between the Broadgate centre in the east of the City, the largest of the 1980s developments which only came under the City's jurisdiction in 1994, and, the ward of Spitalfields, the centre of London's Bangladeshi community and one of the most deprived wards in London, situated directly north-east of the City. He commented that: 'if the Broadgate development epitomises the emergence of London as a 'global city', then the poverty and deprivation experienced literally in its shadow in Spitalfields alert us to the huge social disparities that are increasingly evident in such cities' (ibid. p.2). In subsequent chapters the relationship between the City and the surrounding areas will be developed, and it will be argued that that the addition of anti-terrorist fortification served to reinforce this disconnectedness between the City and the neighbouring boroughs.

In short, *global* enclaves can be seen to embrace inclusion in the globalisation process whist at the same time excluding themselves from the rest of the city through their 'hardened' territorial boundedness. As such, fortified solutions can be seen to most frequently 'help protect and enforce the privileges of social elite areas, and areas of economic investment – the corporate office enclaves and new consumption spaces of the post-modern city' (Graham and Marvin, 1996, p.222).

Today's cities are also characterised by *local* enclaves, which do not have a global function or reference *per se* but which are equally as important to the functioning of the city. Examples of these can include shopping malls, libraries and schools, which, at least in America, are becoming increasingly fortified (Davis, 1990; Lees, 1998). However, perhaps the most noticeable manifestations of local enclaves are so-called gated communities where there is an attempt to create new communities in privately owned and highly defended spaces. For example, Merrifield and Swyngedouw (1997, p.11) indicate how today in advanced western cities new technologies and consumer preference has led to a dramatic increase in gated communities where:

> The powerful...are now able to insulate themselves in hermetically sealed enclaves, where gated communities and sophisticated modes of surveillance are the order of the day. Concurrently the rich and powerful can decant and steer the poor into clearly demarcated zones in the city, where implicit and explicit forms of social control keep them in place.

By the middle of the 1990 it was reported that in many parts of the United States one-third of new communities are incorporating such fortifying principles into their design because 'terrified by crime and worried about property values, Americans are flocking to gated enclaves in what experts call a fundamental reorganisation of community life' (Dillon, 1994, p.8). Furthermore, the Census Bureau's 2001 American Housing survey[6] revealed that over 7 million households (6% of the total) live in gated communities predominantly in sun belt metro areas such as Dallas, Houston and LA, but also that such living arrangements are becoming increasingly popular in places like New York, Chicago and Washington D.C. (see also Blakely *et al*, 1999). This trend is not just related to the United States. Most advanced countries are experiencing this trend to different degrees. For example Caldeira (1996) showed how such developments are altering the physical and cultural landscape of Brazil (see also Schiffer, 2002), and Doeksen (1997) showed similar changes occurring New Zealand and Australia. From a British perspective, Marxist commentator, David Widgery (1991) described what David Harvey (1996) refers to as 'urban apartheid' in London's East End which kept the 'rich' and 'poor' apart due to 'a series of fences, barriers, security gates and keep-out signs which seek to keep the working class away from the new proletarian-free yuppie zones...' (cited in Harvey, 1996, p.409). Such urban areas can be viewed a 'landscape of defence' – 'a landscape shaped or otherwise materially affected by formal or informal defensive strategies to achieve recognisable social, political or cultural goals...[which] may be seen in terms of rich diversity which extends from the loci of violently contested conflict to places heavily invested with symbolic meaning that helps provide a reliable background to everyday life' (Gold and Revill 1999, p.235).

Between them new *global* and *local* enclaves are increasingly becoming reference points in the city, creating a hardened geography of fragmented territories that limit levels of contact with the rest of the urban area. As Sibley (1990, p.6) noted, 'there are implicit rules of exclusion ...that contribute to the structuring of society and space in a way that some will find oppressive and other appalling'. McLaughlin and Muncie (1999, p.120) further highlighted potential 'communitarian-type arguments' for the increased use of urban fortifications. They note that such territorialisation can produce a new sense of localism and that fortifications can stabilise an 'unstable' urban environment. Luymes (1997, p.191) also indicated that such communities are a reaction against the social fragmentation of urban identity, and, that gated communities are the apotheosis of this as they offer a sense of localism and place to their inhabitants, fuelled by the desire for privacy and private property:

> Given the fragmented nature of contemporary urban structure and the suburban culture emphasising localism, privacy and security, it is not surprising that 'closed' subdivisions are becoming commonplace.

In a similar vein, Doeksen (1997) pointed to the construction of such residential fortresses as a fundamental 'expression of distrust in the public domain', more than as a need for privacy. However McLaughlin and Muncie (1999, p.120) countered these arguments by noting the downside to the proliferation of fortified enclaves, notably that it narrows notions of citizenship and governance and can create and sanction 'paranoid discrimination' between those included and those excluded.

In short the impact of both *global* and *local* enclaves can be summarised by noting that all human landscapes should be seen as 'landscapes of exclusion' with the rich and powerful monopolising space excluding the weak to less desirable environments (Sibley, 1995, p.4). This can be expressed by the territorial enclaving of the city in an attempt by social groups to achieve 'spatial purification' (ibid. p.77).

The next chapter will pick up this theme, highlighting how the discourses influencing the proliferation of security measures in the postmodern city are being controlled by certain groups for their own benefit.

Notes

1 It should be noted that in the UK outer-estates for lower socio-economic groups disrupt the generalisation that suburbia was exclusively for the richer sections of society.
2 This idea was also put forward by Jacobs (1961), who saw mixed land use in the city as achieving greater safety as this would increase the times of day the streets were frequented.
3 This was the forerunner of the ring of steel.
4 See *The Belfast Telegraph*, 29 March 1976.
5 Meetings conducted with RUC members December 1995.
6 Cited in *USA Today*, 15 December 2002.

Chapter 3

Controlling the Security Discourse in the Postmodern City

Introduction

In recent years we have witnessed deep-seated changes in the landscape of many Western cities as a result of successive waves of economic, social, political, technological and cultural transformation (Soja, 1989; Harvey, 1990, 1996; Watson and Gibson, 1994; Scott and Soja, 1997). The result of such changes has been that the development of new urban forms has become increasingly profit-driven and security-centred. Such development has also tended focused in the specific parts of the city that the property market will support, further enhancing the fragmentation of the urban landscape (Zukin, 1995; Fainstein, 1994; Pryke, 1994; Graham and Marvin, 2001). As Jewson and MacGregor (1997, p.1) contended, this has led to:

> New patterns of possession and dispossession in urban spaces, the production of cultural representations and city images, the evolution of novel forms of political power, emerging patterns of policing and surveillance, the development of partnerships between public and private agencies, the mobilisation of resistance by urban residents and implications for the empowerment of communities and individuals.

As such, coupled with changes to the physical landscape has been the establishment of new forms of urban governance as the relative role of the public sector in service provision has reduced and the responsibility is increasingly placed in the hands of new forms of public-private partnership often linked to town/city centre management or business improvement districts (Harvey, 1989; Boyle and Hughes, 1995; Gottdiener, 2001).

In particular, increasing attention has been paid to the specific roles played by key urban managers in constructing new forms of institutional arrangements, governance arenas and partnerships, which in turn serve to influence urban form and design solutions linked to the continual securitisation of the city. These different actors have a variety of different motivations which affect what is built, where, and in what style. It is therefore vital to study the complexities and interrelationships between such institutions and how they constrain action, to properly understand the contemporary urban landscape (Wolch and Dear, 1989; Goss, 1992; Jacobs, 1993; Guy and Harris, 1997).

This chapter is concerned with a number of key ideas. It will highlight the extent to which powerful groupings or individuals have sought to control the security discourse that is becoming increasingly influential in urban affairs. It is often noted that the power to shape places is often in the hands of a dominant elite who seek to influence urban change for their own benefits. In recent years there has been an appreciation that the 'density' and 'quality' of the relationships between powerful stakeholders can be of significant importance to the economic position of a particular place. As such powerful economic and political elites often attempt to develop 'institutional capacity' or 'institutional thickness' around key issues (Rose, 1996; Raco, 1999).

This is a central idea within the context of this book, which investigates the role of a number of institutions involved in implementing and constructing anti-terrorist security measures in the City of London. This formed what will be referred to as the inside discourse. This book focuses in particular upon three key institutions – the insurance industry, the City of London Police and the Corporation of London, as well as and outside discourse – a number of more peripheral actors who opposed the enhancement of anti-terrorist security measures in the City for a variety of reasons.

This chapter will also focus upon recent accounts of the fortification of the urban landscape as a result of such a powerful pro-security discourse. The argument is that these discourses have been disproportionately focused on the Los Angeles model, providing a dystopian outlook on future city life. The final section, will then move away from the normative rhetoric created by 'Fortress LA' and analyse the ways in which UK city authorities and agencies of security in the 1990s and early twenty-first century have attempted to control the urban landscape through a combination of management, fortification and surveillance measures as a response to the occurrence and fear of crime and terrorism.

Institutional Capacity and Thickness

The capacity of institutions to react and adapt to change is built up in a set of particular circumstances, often related to 'threats' or 'prizes' and characterised by a strong local political leadership (Wenban-Smith, 1999). Recent work by urban theorists has indicated that a key characteristic of economically successful regions is the ability to change their institutional structures to cope with the changes of a dynamic economy (Harrison, 1994; Cooke, 1995; Lowther, 1999). This is commonly referred to as institutional capacity or thickness. These are terms which are now used almost synonymously to refer to the creation of new networked structures and interactions within a locale by which certain interest groups attempt to assert their influence (Amin and Thrift, 1994, 1995; Healey, 1997, 1998). Institutional capacity/thickness relates to the quality and permanence of locally derived stakeholder networks which build a framework through which shared problems and concerns can be addressed. As a result there is an increasing recognition that institutional capacity matters in relation to achieving the service delivery and planning objectives of the local state (Healey 1998). This framework argues that the ability of a place to achieve economic success is not simply down to a narrow set of economic criteria but that 'the capacity to "pin down" or territorially "embed" global processes in places is becoming increasingly dependent on a whole series of social, cultural and institutional forms and supports' (Macleod and Goodwin, 1999, p.512).

Of particular relevance, given the focus of this enquiry are the institutional arrangements within the City of London that have been highlighted as central to its success as a key financial centre (Sassen, 1991; Pryke, 1991, 1994; Budd, 1998). In particular, research centred specifically on these networks by Amin and Thrift (1994, 1995) attempted to follow similar assumptions made by other institutionalists who attempted to highlight the influence of socio-cultural relations in the economic process (see for example, Granovetter, 1985; Hodgson, 1988; Belussi, 1996). An institutionally 'thick' locale can be characterised by a strong institutional presence, high levels of interaction and the development of a well-defined power and control structure. Amin and Thrift (1994) further noted that institutional thickness, as exemplified in the City of London, was developed by four factors: a high number of civic institutions; high interaction between social groups; coalitions with crossed individual interests; and a strong shared purpose. This work highlighted that socio-cultural factors are significant influences on securing local economic growth with an institutionally thick locale displaying a high level of institutional capacity, which can be seen as 'the ability of place focused stakeholders to... make a difference to the qualities of their place' (Healey, 1998, p.1541).

In this sense the emphasis on the socio-cultural relations that exist within the economic process mean that networking and relational webs are regarded as a capital asset, indicating that institutional capacity is seen as a socio-cultural resource (Suh, 1999). As Raco (1999, p.956) noted:

> Institutional thickness emphasises the importance of local social and cultural relations...It stresses the importance of the local *milieu* – the socio-economic environment of an area resulting from the interaction of firms, institutions and labour, which leads to a common way of perceiving economic and technical problems and finding respective solutions.

In specific relation to the City of London, Pryke and Lee (1995, p.30) further showed how such local *milieu* was central to the financial nexus that has developed in the Square Mile. They noted that 'economic activity is a social and cultural process – not merely shaped and directed by distinctive sets of social relations but constituted through social and cultural practices… that cannot be reduced to inert stimulus-response models'.

It will be highlighted in Part II of this book how institutional thickness was a central feature, which framed the response to terrorism of key institutional stakeholders in the City in a number of ways, noting how institutional arrangements serve to couple local and global relations. First, it will show how new institutional arrangements facilitated new 'interinstitutional arrangements' (Raco, 1998) which can be seen as an attempt to keep the City of London competitive within the global economy, and to remove the negative images surrounding terrorism. Second, it will note how these institutional partnerships were constructed both within the Square Mile and between City representatives, adjoining boroughs and Central Government, in an attempt to reduce the risk from further terrorist attack. Third it will be highlighted that the terrorist attacks and subsequent security responses in the early

1990s occurred at a time when the City was attempting to increasingly promote itself on the world stage, given a series of concerns over its global economic position in the light of development of competing financial areas. Importantly it will show the power of certain dominant 'voices' to influence the make-up of the strategies enacted.

Dominant 'Voices'

Institutional research has sought to illuminate the multiple 'discourses' that exist within the city-making process. In this context a discourse refers simply to the recognisable agendas of individuals and institutions (such as economic development, localism, conservation, transport improvement, cultural promotion and urban renaissance). Commonly these agendas overlap, which can cause the establishment of discourse coalitions or 'communities of interest' (such as pro-development, anti-development, alternative development or, in the case of this book, the defence of the City against terrorism).

Within any locale there will, at a general level, exist what will be termed 'inside' and 'outside' discourses. The inside discourse in this sense refers to the behaviour and decisions of the key actors who are central to developing institutional thickness and capacity. The decisions of the inside discourse are constantly questioned and negotiated by the outside discourse, which is made up of the less powerful institutional voices and pressure groups. Often these 'minority voices' are dismissed, meaning the inside discourse comes to dominate, thus simplifying the planning process considerably by reducing consultation and public protest. What the institutional approach indicates is that collaboration between different discourses, representing different interests, can, and should, occur and create an additional layer of cultural formation or a shared 'frame of reference' (Jacobs, 1994; Phelps and Tewdwr-Jones, 1998; Healey, 1998). For example, Healey (1999, p.1) noted that policy discourses are tied up with institutional work where new relationships are built up and new frames of reference established:

> Policy processes are viewed as a product of complex social relationships through which 'political communities' articulate ideas and frames of reference which guide the way collective resources... and rules... are deployed... These ideas and the frames of reference within which they become embedded (policy discourses) carry power into the fine grain of action (the practices of agency)...[1]

The institutional capacity/thickness perspective tries to show how the development of new institutional structures helps to counter collective concerns and reduce the vulnerability (and hence increase the competitiveness) of local territories. However, what most research in this field fails to note is the danger that certain urban managers are over-represented or 'over-powerful' in the webs of institutional arrangements, leading to domination of local agendas by a dominant institutional elite (inside discourse) which can serve to enhance, rather than reduce, local conflicts, often producing a great deal of mistrust about local politicians and planners and policy makers.

In short, the relationships between institutions are locally contingent and reflect broader patterns of power in a particular place. As Macleod and Goodwin (1999, p.514) pointed out:

> Some institutions are more equal than others when it comes to building and deploying policy agendas…in concrete-complex terms, some institutions posses a 'whip-hand' or are more important than others in helping to construct and cement the thickness.

This they argued often leads to a situation where 'the substantive forms of partnership evolve as a politically 'thin' and mechanically driven strategy' (ibid.).

Shared discourses that develop are more often than not shaped by the actions of powerful actors within the network, who seek to determine the decisions of the policy process as a whole. This means 'power remains largely in the hands of those who make the rules about who can participate and on what terms, and the gatekeepers who allocate resources' (Bailey, 1999). Dominant institutions therefore drive political actions and planning agendas. For example, Beazley *et al* (1997) in a study of downtown development and community resistance highlighted how pro-growth interests subsumed community concerns and that 'powerful forces are at work which can skew democratic decision-making processes in favour of those who control the economic and political life of the city' (ibid. p.191). This they highlighted was 'brutal and socially regressive' and exemplifies the dangers of entrepreneurial modes of governance, which focus on profit maximisation schemes with minimal public consultation in only selected areas of the city in an attempt to boost the economic competitiveness of the city area.

Institutional research, however, also highlights – albeit to a lesser degree – how those stakeholders who contest urban policy can challenge the plans and decisions of the inside discourses, and produce and disseminate new discourses. As Healey (1999, p.1) noted 'policy agendas are reinterpreted and remoulded to create different discourses which have the potential to maintain alternative sources of power and act recursively on the original frames of reference and transform them'.

It will be shown in Part II how a general pro-security 'community of interest' built up within the City of London around the key institutions of the City of London Police, the Corporation of London and the business community (in particular the insurance industry). Their joint concern was to find ways of reducing the physical and financial threat of terrorism. However, it will also be argued that the key managers within this process also had very different ways of pursuing this goal, leading to diverse strategies that were, in some cases, in direct opposition to each other. Furthermore, it will be shown that a number of institutions and organisations outside of the inside nexus of the Square Mile opposed the plans drawn up by the Corporation of London and the police to reduce the threat of further terrorist attack. Like the inside discourses, the outside discourses were fragmented, embodying a number of different arguments why the suggested proposals should not be implemented. This, it will be argued, led to an all to familiar story of the inside discourses marginalising the outside discourses, with the powerful institutions controlling local policy agendas.

Pro-security Discourses and 'Fortress LA'

In recent years, new defensible space approaches to the operationalisation of pro-security discourses are once again serving to influence the design and management of the urban landscape. The response of urban authorities to insecurity has in some cases been dramatic, especially in North America, and in particular Los Angeles (LA) where it is argued that the implementation of crime displacement measures has been taken to an extreme. In LA the social and physical fragmentation of the city is often shown to be very pronounced, and which, according to certain commentators could set a precedent for what it is commonly termed 'postmodern urbanism' (Soja, 1989, 2000; Davis, 1995; Scott and Soja, 1997; Dear and Flusty, 1998; Dear 1999).

During the 1990s LA assumed a theoretical primacy within urban studies with an overemphasis on its militarisation, portraying the city as an urban laboratory for anti-crime measures (Davis 1990, 1992, 1998; Flusty, 1994; Christopherson, 1995; Crawford, 1995). Fortress urbanism was highlighted as the order of the day, as an obsession with security became manifested in the urban landscape with 'the physical form of the city… divided into fortified cells of affluence and places of terror where police battle the criminalized poor' (Dear and Flusty, 1999, p.57). For example, it was reported that in 1991 that sixteen per cent of Los Angelian's were living in 'some form of secured access environment' (Blakely and Snyder (1995, p.1).

Mike Davis is perhaps the most cited author on 'Fortress LA'. Davis (1990, 1992, 1998) depicts how in recent years the authorities and private citizen groups in LA have responded to the increased fear of crime by 'militarising' the urban landscape. His dystopian portrayal of LA in *City of Quartz* (1990) provided an alarming indictment of how increasing crime trends could theoretically affect the development and functioning of the future city through the radicalising of territorial defensive measures with the Los Angeles Police Department (LAPD) becoming a key player in the development process (see also Herbert, 1996). As the boundaries between the two traditional methods of crime prevention – law enforcement and fortification – have become blurred, defensible space, once used at a micro-scale level, was being used at a meso and macro level to protect an ever-increasing number of city properties and residences.

In *Beyond Blade Runner – Urban Control, the ecology of fear*,[2] Davis (1992/1998) extrapolated current social, economic and political trends to create a vision for the future city in the year 2019, which in this account had become technologically and physically segregated into zones of protection such as high security financial districts and segregated gated communities. In this vision, economic disparities have created an urban landscape of cages and wasteland.[3] He argued that defensive strategies already used in LA need only be augmented to 'perfect' this vision.

In this account Davis portrayed fear and anxiety in LA as symptomatic of the negative consequences of the rise of capitalism. Davis's ideas are made more explicit in Dear and Flusty's (1998) account of postmodern urbanism, which saw the globalisation of the economy as the driving force and which created a series of local social, political and economic inequalities which are reflected in the fragmentary and fortified urban form. In this scenario, as the educated and mobile residents sought the 'dreamscapes' of privatised suburbia, the poor and ill-educated remained in the 'carceral cities'. Other commentators have taken up this theme noting that 'as some writers suggest Los Angeles is a crystal ball of capitalism's future, the forlorn dystopia of *Blade Runner* is maybe just around the corner' (Merrifield and Swyngedouw, 1997, p.12).

Davis's work was also elaborated on by Flusty (1994) who provided a categorisation of the different types of fortress urbanism, which he argued, had thrown a blanket of fortified and surveillance security over the entire city. He referred to the spaces of security as 'interdictory space', which are designed to exclude by their function and 'cognitive sensibilities'. A typology of such spaces is shown below in Table 3.1.

Table 3.1 Typology of interdictory space

Stealthy space	Passively aggressive with space concealed by intervening objects
Slippery space	Space that can only be reached by means of interrupted approaches
Crusty space	Confrontational space surrounded by walls and checkpoints
Prickly space	Areas or objects designed to exclude the unwanted such as unsittable benches in areas with no shade
Jittery space	Space saturated with surveillance devices

Source: Adapted from Flusty, 1994.

Flusty highlighted how such defended spaces, alone or in combination, have pervaded all aspects of urban life, leading to an ever increasing number of highly secure gated communities, bunker architecture and highly policed ghettos in the disadvantaged poor areas of the city. He also noted how the commercial privatisation of space is taken to an extreme as a strong fear of the public realm leads to highly inclusive business facilities either in isolation or in self-contained agglomerations.

Although 'Fortress LA' has become a powerful vision for the city it is important to realise that there are many ways in which urbanism in LA may be viewed. Critics of Davis, for example, argue that he 'contends that nature makes Los Angeles the most dangerous city in America' (Friedman, 1998) and that he 'is selling fear and anxiety about LA' (Stewart, 1998). However, not all images of LA are negative (Benham, 1993), although most point to the potential for crisis. Benton for example

(1995, p.145) notes that the same urban landscape in LA can depict a variety of kaleidoscopic images and landscape meanings:

> Turned one way the kaleidoscope reveals images of a romantic, utopian Los Angeles. Turned another way, it shows a landscape of fear, inequality, and violence and the struggle to survive. Twisting the kaleidoscope once again, Los Angeles becomes a landscape of impending conflict and convergence, a city swirling with apprehension. All these representations could be accurate.

Postmodern Urbanism

In recent years, some urban theorists have argued that notions of the 'fortress city' can be seen as part of the wider attempt to provide a model of what future cities will look like under the banner of postmodern urbanism.[4] In contrast, despite the apparent omnipresence of fear and defensive landscapes in LA, others have highlighted that it is important not to generalise about these trends as they appear to be very specific to this particular city (Hall, 1966; Merrifield, 1997; Oc and Tiesdell, 1997).

LA has been viewed for most of its history as atypical of the American urban area, but that there is a growing speculation that it *could* provide the prototype of the 21st century American City (Suisman, 1989; Dear, 1995; Scott and Soja, 1997). As such, LA in the 1990s assumed a position within urban studies similar to the position Chicago held in the 1920s – that is as a city which, some argue, exhibits particularly stark urbanising processes which generally can be found throughout much of the global economy (Savage and Warde, 1993). In the previous decade the so-called 'California school' or 'Los Angeles school' have made important contributions to contemporary urban theory.[5] These often Marxist-based accounts implicitly argue that changes in the morphology of LA are related to the changing structure of global markets and that this throws up an urban geography with particular emphasis on the negative effects of capitalism such as social exclusion and increased economic polarisation between rich and poor.

In recent years the most noted work on contemporary LA *per se* has been undertaken by Michael Dear and others (Dear and Flusty, 1998, 1999; Dear, 1999) who pose the questions 'have we arrived at a radical break in the way cities are developing? Is there something called *postmodern urbanism*, which presumes we can identify some form of template that defines its critical dimensions?' (Dear and Flusty, 1998, p.50). They argued that a new theory, with its own language and set of propositions is required to describe these new urban forms (see also Soja, 2000).

Dear and Flusty (1998) were quick to dismiss the work of the Chicago school, noting that much of the urban research agenda of the twentieth century has been predicted on the concepts of the concentric zone, sector and multiply nuclei theories of urban structure (see for example Burgess, 1925; Park, 1929; Hoyt, 1939). They argue that the work of the Chicago school was based on a modernist perspective of external gaze and detachment. However, as a number of critics have noted, this does not do justice to the Chicago school, failing to account for the rich ethnographic work that was involved in constructing the basis for the structural models (Linder,

1998; Beauragard, 1999; Jackson, 1999; Lake, 1999). Dear and Flusty (1998, p.53) also made a similar criticism about previous attempts to study LA (see for example Banham, 1973; Soja, 1989; Suisman, 1989), noting that these commentators 'maintained a studied detachment from the city... through a voyeuristic, top-down perspective'.

Despite attempting to move away from this detached perspective to highlight the multiple socio-cultural, economic and political geographers within LA through 'postmodern sensibilities', their focus is very much based on urban structure, with little indication given to the multiplicity of 'voices' that could be found. As Lake (1999, p.394) noted, 'in place of a self reflective internal unfolding [they] offer the same God's-eye view of Los Angeles from an external vantage point as that inscribed in the Chicago models'. Jackson (1999, p.400) was more critical, arguing that Dear and Flusty's account was 'ethnographically void' and that they 'simply "read off" a series of meaning from the urban environment, leading to an exaggeration of the power of academic interpretation and diminishing the agency of "ordinary people"... the actual voice of those who live in Los Angeles (besides a handful of academic theorists and planners) go unheard and unrecorded... Instead their prose is populated with the cool abstractions of social polarisation and fragmentation'.

The morphological emphasis that pervades Dear and Flusty's ideas was seen to be a result of their failure to question the Chicago schools emphasis on urban form which leads to their research being predominantly articulated in terms of urban structure. This in turn de-emphasises the key processes central to the creation of urban landscapes (Beauragard, 1999). Dear and Flusty argued that what they are studying is urbanism, but as Jackson (1999) noted, the Chicago school always saw '*Urbanism as a way of life*' (Wirth, 1938) and not simply as urban structure or morphology, but containing a rich mix of human experience. Jackson (1999) argued that it is perhaps through an enhancement of the ethnographic dimension that we might be able to see clearer the significance of contemporary LA for urban process in general, thus avoiding the 'fanciful caricature and superficiality' of previous accounts of LA (Soja and Scott, 1986, p.249, cited in Jackson, 1999, p.402).

In short the LA model of contemporary urbanism has a tendency to reduce urbanisation to economic causality and associated urban forms, which are too extreme to become generalisable (Savage and Warde, 1993). This breaks with another stream of current thinking in urban theory, which is 'less intent on parsimonious descriptions and abstract and general explanations than on evoking rich images of a single city' (Beauregard, 1999, p.398). As Sassen (2000, p.144) further noted:

> It is perhaps one of the ironies at this century's end that some of the old questions of the early Chicago School of Urban Sociology should re-emerge as promising and strategic to understand certain critical issues today, notably the importance of recovering place and undertaking ethnographies at a time when the dominant forces of globalisation and telecommunications seem to signal that place and the details of the local no longer matter.

Relating the LA Experience to British Cities

As noted above recent writings about cities often present the complexity of these contemporary urban transformations as a series of broad general trends associated with postmodern urbanism. Whilst LA is often held up as a template of contemporary or future urbanism, and in particular the Fortress LA model, it can be argued that the universal applicability that is implicit in such accounts is misguided. This can be seen as part of a much larger normative debate within the social sciences brought about by the 'conception of a break', which pervades much of the contemporary literature on western society (Healey *et al*, 1995, p.6). This includes the idea of an epochal shift between modernism and postmodernism, between forms of economic organisation (Fordist and post-fordist) between new types of society (industrial and post industrial) and how that society deals with new forms of global risk (industrial society to a risk society – see Chapter 4). As Gold and Coaffee (1998, p.286) noted in relation to urban geography:

> The research literature struggles to identify elements and processes that transcend specific circumstances of individual case-studies. Pedagogic literature that adequately surveys the complex realities of the current urban scene is hard to find... [and] the models and frameworks by which we previously understood the city have been undermined without the creation of adequate replacements. Even basic definitions remain contested.

Although the fortification tendencies in LA are far more dramatic than that which has so far occurred in Britain, the prevalence of defensive landscapes in the central business district and in middle class residential communities are increasing trends in many cities in the UK (Graham and Marvin, 1996, 2001; Oc and Tiesdell, 1997; Fyfe and Bannister, 1998).

We should however be careful about including extreme urban experiences when hypothesising about the future UK city. Peter Hall (1966, 1977) in *The World Cities* first highlighted the dangers of inferring trends prevalent in LA will occur in the UK. In a similar way the writings of the self proclaimed LA or California Schools on 'Fortress LA' should not however be taken as a model for other cities, especially those in Britain. As Merrifield (1997) argued, 'do British-style urban trends for instance, really chime with some of the American (Californian) hyperbole?' (p.59). He continued:

> Are our cities being turned into theme parks or into the dystopian horrors of *Blade Runner*? I don't think they are. This isn't to say that piecemeal commercialization and McDonaldization hasn't taken place nor that CCTV surveillance hasn't proliferated in shopping centres, in residential complexes and in strategic centres of power... Nor does it deny how theses developments have enabled the wealthy to reinforce their power and control over urban space and marginalize the poor and homeless into their own distinctive enclaves (such as assorted Skid Rows, deserted parks and unattended doorways).

Oc and Tiesdell (1997, p.194-5) also indicated that British cities could avoid such fortification as exemplified in LA by rejuvenating central city areas:

> Los Angeles' 'fortress downtown', as colourfully depicted by Davis, is probably a singular case where extreme measurements were needed due to the extremity of the city's problems. If the necessary steps are taken in Britain to reverse the decline of city centres then British cities will not have to resort to creating similar fortress city centres.

It is therefore important not to let the reported situation in LA (and the United States generally) influence the ways in which we view the British city. As Merrifield (1997) reiterates it is important that 'the hyperbole and the prophesies of doom and dystopia coming from the United States aren't simply used as sloppy catch-all categories or simply mapped on to the British experience uncritically'. Undoubtedly, defensive strategies employed in the city are becoming more widespread, especially given the recent emphasis on urban renaissance and city centre living (Urban Task Force, 1999). However, the balance has to be struck between, on the one hand, fortifying the landscape so that it actively excludes people, and on the other hand, providing adequate security so that fear of crime is reduced and citizens give the central areas of cities patronage. Examples given in the last chapter from central Belfast and in this chapter from LA highlight extreme examples of how, if used to predominately exclude, strategies of urban security can destroy the public realm and severely influence the marketability of particular cities.

Part II of this book will highlight these ideas in relation to the City of London's attempts to reduce the risk of terrorist attack through the 1990s and subsequently in the early years of the twenty-first century. It will show that those responsible for security in the City had to consider whether they wanted to create a fortified citadel in the vision of 1970s Belfast or LA in the 1990s, or construct a subtler regime of defensive alteration that balanced the needs of security with business continuity.

The next section of this chapter highlights the three overarching and intertwined strategies used by the agencies of security and in UK cities to improve safety and reduce the fear of crime. This section will also note how these strategies are managed, as well as the potential negative side effects of such securitisation in terms of civil liberty concerns and the displacement of criminal activity.

Strategies of Urban Defence

Today a number of general strategies to deal with city security can be noted: first, the *management* of the landscape where a series of spatial and temporal rules and regulations are socially enforced or dictated by the forces of law and order or other key urban managers; second, the *fortification* approach that refers to the privatisation of space due to the introduction of defensive measures such as walls, barriers, and gates, which causes physical segregation of the landscape; third, the *surveillance* approach referring to the control of space through an explicit use of security cameras or security personnel, particularly at the entry into a territory.

The Management of the Landscape

The managerial approach calls for the implementation of greater legal controls with greater numbers of police who have more power to shape the way in which urban landscapes and community life are structured (Rubinstein, 1973; Grogger and Weatherford, 1995; Wekerle and Whitzman, 1995; Mitchell, 1997). For example Neil Smith (1996) showed how the homeless (who were seen as the intruders) were systematically evicted from a park in New York due to moves by the police, and the new middle class residents, to upgrade and defend the character of the area against those seen as 'dangerous' (see also Brantingham and Brantingham, 1984, 1990; Zukin, 1995; De Souza, 1995; Fyfe, 1997). However, perhaps the work which has received most attention in regard to how the police strategically use notion of territoriality to control area of the city is *Policing Space*, Steve Herbert's ethnographic study of the LAPD (see Herbert 1996, 1997a, 1997b, 1998). This study indicated the significance of territoriality for the LAPD who seek to control the spaces they patrol. The importance of territoriality, states Herbert (1997b, p.399), intensifies as power is resisted – 'contested spaces preoccupy the police most'. It highlights how police strategies involve creating boundaries and restricting access as they seek to regulate space.

Herbert's work, despite being criticised for its lack of socio-economic context which effects the policing regime in this particular place (Fyfe, 1997; Marston, 1997), does provide insights of general applicability to this study, given that one of the key instigators of the anti-terrorist security measures in the Square Mile were the City of London Police. First, it highlights how the police are now increasingly active in the planning process and strive to form partnerships with local authorities or community groups, to reduce crime and the fear of crime in the public realm. Second, it highlights how the actions of the police can be interpreted through the control of space. Third, it shows how the 'power of the police is inserted within the fabric of the city' (Herbert, 1997a, p.6/7).

In addition to the work of the police, the private security and risk management industries have grown at a fast rate over the last twenty years, given rising crime rates and an increased fear of crime (Ericson and Haggerty, 1997; Loader, 1997; Griffith, 1999). There are now a multiplicity of agencies that undertake policing functions, other than the police service. For example, Jones and Newburn (1998) indicated that the number of people working in private security now approximates to that of the 43 police forces in England and Wales. They also showed how there tends to exist a 'benign co-existence' (ibid. p.181) between the local police force and the private security agencies who tend to operate in different geographical spaces – the police in the public realm and the private security professionals in the spaces of commerce.

In Part II it will be shown how the police instigated a series of anti-terrorism procedures in the City of London, but, that they were backed-up by private security personnel and risk management specialists, in the management of the secure zone.

In addition, the actions of both the police and the private security industry have been aided in recent years by a host of specific legalisation linked to countering the terrorist threat. For example in the UK we have recently seen:

- The Prevention of Terrorism (Additional Powers Act) 1996, which backed up previous legalisation (1974) and further allowed for random police searches, and for the imposition of police cordons. This act was rushed through Parliament in April 1996 due to fears of further Provisional IRA bomb attacks following the large explosions in the London.
- The Terrorism Act 2000, which came into force in February 2001. This act provided a permanent UK-wide anti-terrorist legislation to replace a number of separate pieces of legislation. It sought to redefine and in essence broaden definitions of what was considered 'terrorism'. However, the act was criticised as many pointed out that it could potentially turn activist movements into terrorist organisation.
- The Anti-Terrorism, Crime and Security Act 2001 was brought in after 9/11 as a mechanism to tackle new forms of global terrorism especially linked to weapons of mass destruction. It aimed to strike a balance between respecting civil liberties and bringing in targeted measures which could enforcement and intelligence gathering capabilities of the security services. The Act significant extended Police powers for example on data gathering and surveillance, holding suspected terrorists without charge and on seizing or 'freezing' suspected terrorist funds. Like the 2000 act it has been viewed as overly draconian by civil libertarians (Liberty, 2002).

As will be highlighted later in the book these acts although undoubtedly useful for improving security in and around the City of London, have also been used for purposes not directly linked to terrorist activity.

The Fortification of the City

The agencies of security noted above are primarily responsible for the 'fortification approach', which has taken the form of turning office blocks, shopping centres and residential communities into territorial enclaves through methods of restricted access and electronic surveillance. The increase in this type of fortified landscape has led to the increased privatisation of the city, and to what have previously been referred to as global and local enclaves, where there is 'replacement of public access with private spaces that can be controlled by security guards and the ability to pay' (Wekerle and Whitzman, 1995, p.6). This has led to further restructuring of the city 'creating a patchwork quilt of private buildings and privately appropriated space' (Trancik, 1986).

The desire for fortified territories in the city has meant that Oscar Newman and his *Defensible Space* principles (as detailed in Chapter 2) are now back on the public policy agenda in an attempt to make residential communities in America more 'desirable' by altering the design of areas (see for example Cisneros, 1995; Ekbolm, 1995; Newman, 1995, 1996, 1997; Harvey, 1997; Blakely and Snyder, 1997). As Brown (1995) indicated, 'barricades and bollards have become the newest accessory on this country's psychic frontier...You might call it the architecture of paranoia.

They call it "defensible space"'. For example Ellin (1996) noted that Newman had recently been given a grant from the United States Justice Department to improve security in fifty residential areas. Newman himself (1995, p.151) believed crime would be reduced in these areas by 'limiting access and egress to one opening...it was reasoned that such a street system would be perceived by criminals and their clientele as too risky to do business in'. Indeed, findings from his current work show that crime has been reduced by 25 per cent and violent crime by 50 per cent (Newman, 1997). More recently work in the UK by Cozens *et al* (1999, 2000) undertook to test aspects of Newman's theory and found that the design of residential environments was important to the 'image' of the area as well as its perceived 'criminogenic potential'.

Other research results on defensible space principles have differed somewhat, highlighting that residential street closures or traffic modification, while not likely to have a major impact on crime rates, will significantly reduce the fear of crime (Griffiths, 1995; Wagner, 1997). Indeed, tackling the fear of crime is now becoming as important as tackling crime *per se* with the fear of crime perhaps having greater potential to destroy urban communities (Box, 1988; Pain, 1995).[6] For example, recent figures released by the Canada Safety Council in July 1998 indicated that whilst crime rates might fall, fear of crime is still rising due primarily to media 'hype'. The overall crime rate in this instance fell in 1997 for the sixth year in a row, but the same report points out the fear of crime is still on the increase. It is pointed out that 'Canada is not becoming a more dangerous place to live' rather 'it is a myth – fuelled by political expediency, emotion and media hype – that crime is on the rise'.

However, some research views anti-crime fortification measures as ineffective *per se*. Ellin (1997) for example, edited a series of contemporary essays entitled *Architecture of Fear*, which examined 'the ways in which the contemporary urban landscape is shaped by a preoccupation with fear' as apparent in design, security systems, gated communities, semi-public places, and zoning regulations. She noted that such design often acts as a placebo:

> This fixation [with security] manifests itself in such efforts... despite the evidence that they do not lessen crime... [and] that such disjointed efforts exacerbate rather than eradicate the sources of fear and insecurity.[7]

Furthermore, Marcuse (1993, p.101) argued how city walls can be seen as both 'walls of fear' and 'walls of support', whilst Ellin (1997) herself cited that whilst 'form follows fear' in the city this relationship can be reversed to one in which fear is seen to follow form. This infers that changes to the urban fabric intended to reduce risk and crime actually serve to exacerbate the fear of crime. As she noted, 'certainly, the gates, policing and other surveillance systems [and] defensive architecture... do contribute to giving people a greater sense of security. But such settings no doubt also contribute to accentuating fear by increasing paranoia and distrust among people' (Ellin, 1996, p.153).

This relationship between fear and urban form will be investigated in Part II in relation to the anti-terrorist measures constructed by the City of London Police.

Furthermore the overall impact of 9/11 on urban form will be highlighted noting that this event appears to be advancing the barricading of the city leading to increasing partitioning and citadelization of the urban landscape (Marcuse, 2000a).

The Surveillance Approach

The fortification approach is often complemented by enhanced surveillance, in particular from closed circuit television (CCTV), which it is argued can deter crime from areas in which it operates. In the UK the first centralised CCTV scheme was erected in Bournemouth in 1985 to stop vandalism along the sea front. Following this, a series of terrorist bombs, football hooliganism and rises in city centre crime rates encouraged many more local authorities and private businesses to install security cameras (Brown, 1995). Horne (1996), arguing for more CCTV in our cities, indicated that 'the demand for CCTV systems has been in response to increasing crime and incivilities which affect the quality of life' in cities due to its proposed benefits – deterring crime, the freeing-up of police manpower (which can then be redeployed), possible insurance discounts which could be given, and reducing the fear of crime. He further indicated that it will reduce the perception of insecurity, as citizens will feel that they are in a 'protected area'. Some reports indicate that reducing the fear of crime is CCTV's main benefit. For example, the Chief Inspector of Merseyside Police in 1996 indicated that whilst the police will not make any claims about the effectiveness of the CCTV in reducing crime, they are convinced that it has significantly reduced the fear of crime in the city centre, especially at night.[8]

CCTV is now the most common preventative measure taken to stop crime and 'has had more of an impact on the evolution of law enforcement policy than just about any technological initiative in the last two decades' (Davies, 1996b, p.328). For example Fyfe and Bannister (1996/8) noted that over ninety towns in the UK at that time had centralised CCTV systems. This number is rapidly increasing with over 280 further towns considering introducing similar schemes (Poole and Williams, 1996). This meant that in 1996 between £150-300 million per annum being spent on CCTV cameras, which equates to 200,000 cameras, many of which are erected in high-rent commercial areas (Davies, 1996b, p.328). Williams and Johnstone (2000) highlighted that this represented a 550% increase of CCTV in commercial centres between 1994 and 1999. Post 1999 there has once again been a central Government programme (The Crime Reduction Programme CCTV initiative) aimed at enhancing CCTV in towns and cities as a result of attempts to re-invigorate central shopping areas. In short between 1992 and 2002 it is estimated that over £3 billion has been spent on CCTV installation and maintenance in the UK (McCahill and Norris 2002).

It is not just the apparent effectiveness of CCTV in recording crime that is highlighted. The visual deterrents that such schemes produce are often seen as a key function. Geake (1993) for example, cited a security consultant who stated 'the effect of CCTV is 95% deterrent and 5% detection'.[9] Warning signs that are commonly displayed indicating that the area is under CCTV surveillance further reinforce the deterrent value of CCTV. However, Home Office Research conducted in 1995, clouded this perspective showing that the deterrent effect became less significant as time progressed, as criminals began to know the direction the CCTV cameras faced (Ditton 1996, Millward, 1996).

Amongst the agencies of security, the general consensus is that CCTV is a panacea having positive benefits for an urban environment making it feel more secure, both in terms of limiting crime and reducing the fear of crime. Durham (1995) writing in *Police Review* gave an account of the success of a typical CCTV scheme in Newcastle indicating that the system has been a revelation in policing. Starting in 1992, the effectiveness of the scheme can be seen in the 50 percent reduction in overall crime in its first two years of installation. The scheme was a joint initiative between the police, the local authority and local businesses. He also noted that non-CCTV areas also benefit, as resources can be increasingly deployed in these areas to minimise any possible displacement effects. In addition, it is reported that people were now beginning to perceive central Newcastle as a safer place to visit and work. This was reflected in insurance discounts that were offered to businesses located within the CCTV area.[10]

This type of argument has often been supported by the results of surveys carried out by the Home Office and the police, which support the implementation of centralised CCTV systems. The Home Office for example made £5 million available through a 1994/5 initiative (Home Office, 1994) for installation of such camera networks. Additional financing in the following years has also been forthcoming (Horne, 1996; Williams and Johnstone, 2000). Furthermore, research carried out by the Police Authority for Northern Ireland prior to the activation of the Belfast CCTV scheme in December 1995 indicated that the public strongly supported the scheme. From nearly 1500 full and partial interviews conducted 89% of respondents were broadly in favour of the scheme with only 7% against it. Of further note, 98% of those over the age of 65 were in favour of the scheme and 83% thought that CCTV did *not* represent and infringement of their personnel freedom.[11]

However, many commentators are now pointing out what they consider to be the negative impacts of CCTV relating to how it reorders and controls urban life (Lyon, 1994; Sorkin, 1995; Soja, 2000). Such accounts have drawn attention to the growth of surveillance technologies viewing it pessimistically where surveillance signifies social control (Marx, 1985; Lyon, 1994; Davies, 1996a; Norris and Armstrong, 1999). The intense surveillance in certain areas of the urban landscape has meant that the city in this context is viewed by some commentators as now no more than 'a carceral city', a collection of surveillant nodes designed to impose a particular model of conduct and disciplinary adherence on its inhabitants' (Soja, 1995, p.25). The plethora of surveillance technologies within the city are seen to evoke fears of an Orwellian society based on 'Big Brother', or rather a variety of 'Little Brothers' (Lyon, 1994, p.53).

Centralised CCTV systems are often equated with Jeremy Bentham's Panoptican prison idea first formulated in 1791, and seen as a metaphor for control of urban space (Dandeker, 1990; Lyon, 1994; Bosovic, 1995; Marx, 1995; Fyfe and Bannister, 1996).[12] This fully encompasses the notion of territorial control by creating the impression of omnipresence acting as a visual deterrent. As Davies (1996, p.17) stated, 'CCTV... creates a means of enforcing public order on an unprecedented scale'. The panopticon concept has now been extended beyond the confines of individual buildings and into the public realm in an attempt to control urban space (Oc and Tiesdell, 1997).

Indeed, Steve Graham writing in 1999 argued that CCTV was fast becoming 'the fifth utility' – an integral part of the infrastructure of our cities alongside, water, gas, electricity and ICT networks (Graham, 1999; see also Norris and Armstrong, 1999; Johnston, 2002). Perhaps most notably, CCTV has become integrated within many traffic management systems. For example, the CCTV systems around, and within, the City of London, which will be described in detail in subsequent chapters, was one of the first such systems to use digital CCTV technology to automatically read the number plates of vehicles entering or exiting particular areas. Automated Number Plate Recognition systems (ANPR) are now highly advanced and reliable in terms of identification rates (McCahill and Norris, 2002). Today, the technology backing up CCTV is ever advancing especially around biometric technology and in particular the ability to identify facial features (Lyon, 2002). Such systems can instantly compare an image of a face with a database of suspected terrorists or anyone else. For example, a system of 100 biometric cameras was reportedly suggested for Times Square, New York, in the aftermath of 9/11, to scan the faces of pedestrians and then to compare these to a data base of suspects (Rosen, 2001). As will be detailed, similar suggestions were also made in relation to updating the CCTV networks in the City of London.

Civil libertarians have also been worried about issues of accountability and monitoring of such schemes. For example a Home Office survey in the early 1990s (Honess and Charman, 1992) indicated that over 50% of respondents felt neither the government nor private businesses should be allowed to install CCTV without public consultation; 72% thought the CCTV cameras could be abused by the wrong people; 39% had a distrust of the CCTV system *per se*; and perhaps most importantly, 37% felt that in the future there was a danger that the system could be used by the government to control people. Subsequently the Local Government Information Unit (1996) drew up a code of practice for the installation and use of CCTV.

Others have highlighted the potential negative impact of CCTV on police procedure meaning that alternative community methods of crime prevention are often seen as secondary, as CCTV is 'waved aloft by police and politicians as if it were a technological Holy Grail, and its promises chanted like a mantra as a primary solution for urban dysfunction' (Davies, 1996, p.328). Indeed it is commonly highlighted that 80-90% of town centre users support the introduction of CCTV (McCahill and Norris, 2002). As will be highlighted in Part II there were prolonged civil liberty protests about the enhanced use of CCTV in the City of London that formed part of the 'outside discourse', which did not fully support the counter-terrorist strategies that were being implemented in the Square Mile.

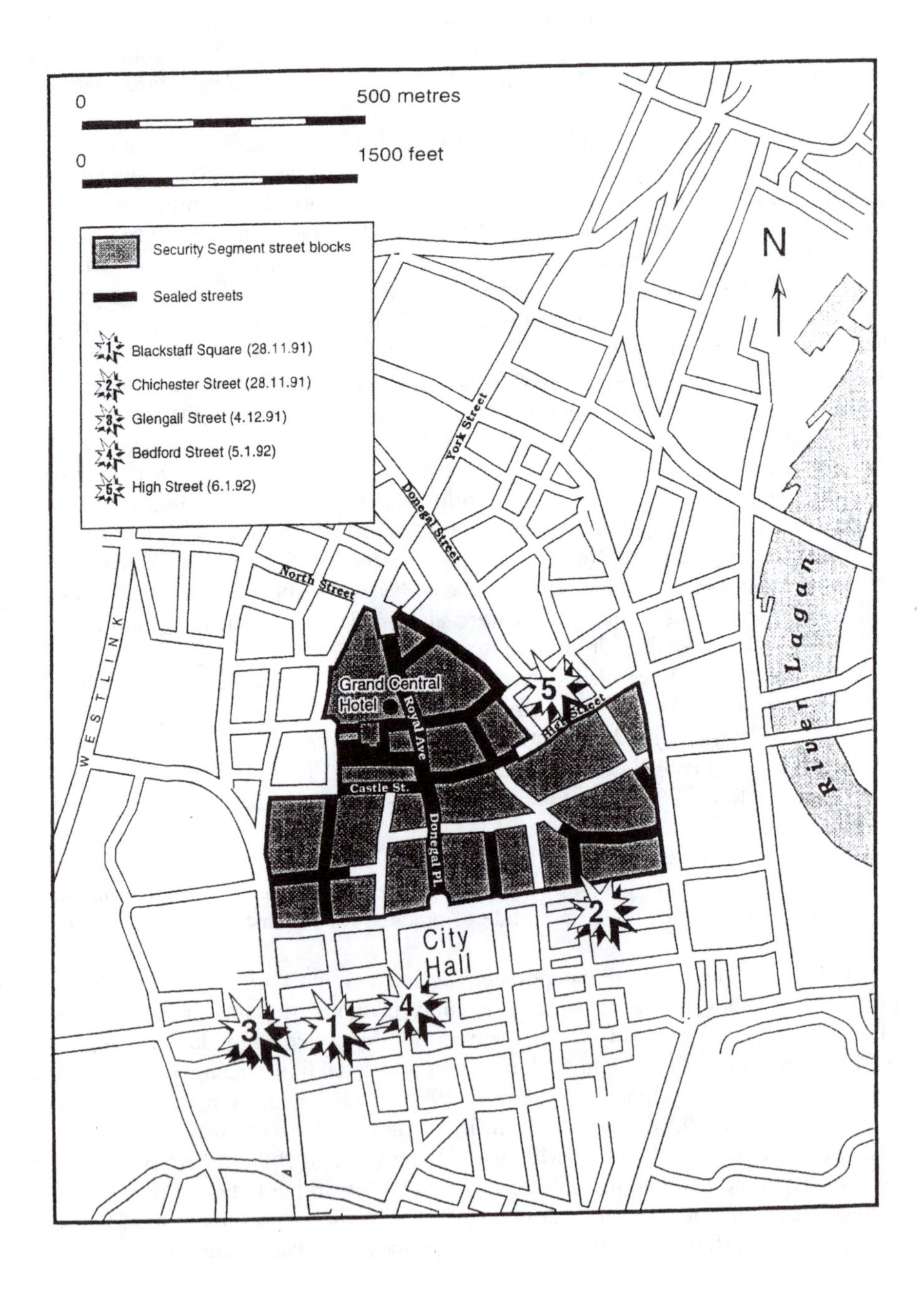

Figure 3.1 The vulnerability of the exterior of the ring of steel: the Christmas/New Year bombing campaign 1991/2

Displacing the Threat of Crime

As defensive landscapes *per se* are becoming increasingly widespread in the city it is important to note how the apparently secure spaces relate to other places within the same city. In particular, the danger of defending one area within a city is that geographical displacement of crime to alternative targets and areas that are not so well defended will occur (Reppetto, 1976; Cornish, 1987; McDowell, 1999).

Barr and Pease (1990) drawing on the work of Hakin and Rengert (1981), suggest that a number of types of displacement can occur: spatial (an alternative location for crime is sought); target (an alternative target sought); or crime displacement (an alternative crime is undertaken in the same locality). Furthermore, Evans (1995, p.95) suggested that it is possible to identity conditions under which displacement could occur. These include the motivation of the offender, the availability of an alternative targets and the location of low vulnerability targets in close proximity to highly vulnerable targets. By contrast, some commentators have noted that it is difficult to prove that such displacement does occur (Gabor, 1981, 1990), whilst others have highlighted that crime prevention measures initiated to stop a certain type of crime often serve to reduce other forms of criminal behaviour in that area (Clarke, 1992). Displacement of terrorist crime forms a key aspect security in the Square Mile. As will be shown, as the City of London has introduced anti-terrorist measures the risk has increased at other sites in the capital. Additionally other non-terrorist related crimes in the Square Mile have been significantly reduced.

The idea of crime transference will be highlighted in Part II of this book. Drawing on experiences from Belfast, it will note how displacement of terrorist activity to the exterior of the ring of security measures encircling the Square Mile was feared in the City, which subsequently led to further security measures being deployed to counter this threat in the areas directly outside the security cordon. Incidents in Belfast such as the Christmas and New Year's bombing campaign of 1991-2 show how the risk of terrorism was exported to the periphery of the security zone, where bombs could be left with less fear of detection and without having to pass through security checks (see Figure 3.1).

Conclusion

With the development of new urban enclaves, cities are becoming socially and spatially restructured with a new territorial order controlling and organising space. In these enclaves, in which form is increasingly following fear, access control and boundary reinforcement are paramount and access and egress regulated. As Oc and Tiesdell (1997, p.16) noted, these:

> New territories in the central city areas are increasingly privatised as the agencies responsible for their creation seek to create a sanitised environment with certain well-controlled and regulated safe areas or spots.

In general, crime prevention measures can serve to enhance social segregation through the construction of defended territories. As Harvey (1996, p.209) argues, the territorial imperative is of considerable importance in the study of place and community and the construction of the geographical landscape, indicating that the fear of the 'other' 'now lead to all sorts of exclusionary territorial behaviour' in the city. Furthermore, Soja (1989, p.150) noted contemporary notions of territoriality refer to 'the production and reproduction of spatial enclosures that not only concentrate interaction…but also intensify and enforce its boundedness'. This has led to the growth of what can be referred to as 'pseudo-public spaces' where commerce has sought to secure their agglomeration of interests by restricting entry to a given area by means of street barriers (Davis, 1990, p.226). What was considered public space is now increasingly enclosed for private benefit. In today's cities there is an emphasis upon the 'jostling of landscapes of consumption, spectacle or power with landscapes of despair, exclusion and negation in our cities' (Badcock, 1996, p.92).

Defending the city is not a new trend and is as old as urbanisation itself. Today's cities have their own expressions of security forged by a series of managerial, fortressing and surveillance approaches. Strategies to design out crime from the early 1970s have been resurrected and images of fortified security are now commonplace and in some cases have expanded exponentially after 9/11. Indeed, the rhetoric of 'the appearance of being safe is almost as important as being safe' has never been more pertinent.[13] Marcuse (2002a) for example points to a number of potential consequences of 9/11 linked to the '*Partitioning of urban space*' ranging from the impact of 'concentrated decentralisation' on the clustering of financial services in certain districts to the increased use of 'citadelisation' – the aggressive fortification of business and residential space' (p.274) and the 'barricading of the city' (p.275) leading to an increase in surveillance and control of public spaces. In short, he notes 'security becomes the justification for measures that threaten the core of the urban social and political life, from the physical barricading of space to the social barricading of democratic activity' (p.276).

In Part II of this book these ideas will be articulated through the case study of the terrorist threat to the City of London in the 1990s and early twenty-first century.

Notes

1 A more detailed analysis of discourse analysis lies beyond the remit of this book. For a fuller explanation see Fairclough, 1992; Burr, 1995; Richardson, 1996.

2 The bleak landscape of LA has provided the backdrop to many futuristic films, with perhaps Ridley Scott's *Blade Runner* providing the best example. In this film urban decay, as characterised by crime, terror and drug-ridden environments, are found side by side with the powerful secure corporate office fortresses from which LA 2019 is run.

3 In Mike Davis's proscription there are perhaps three types of area in LA that have most undergone such militarisation. The *first* is the core area as represented by the high security financial complex of Bunker Hill. In this zone the police, according to Davis, have become central development players as the whole area has been put under the constant gaze of

electronic surveillance. The *second* defensive zone of the city that Davis highlights are the 'gated communities' patrolled by armed security guards, which effectively segregate these areas from the rest of the city. The *third* secure zone Davis mentions are what he calls 'urban simulators'. With tourists unwilling to venture into what they believe to be the dangerous parts of LA, the Disney Company and others have recreated 'vital bits of the city within the secure confines of fortress hotels and walled theme parks'.

4 To discuss in detail the arguments surrounding the postmodern city see Hassan, 1985; Cooke, 1988; Featherstone, 1988; Best and Kelner, 1991; Bradbury, 1995; Cuthbert, 1995; Kumar, 1995; Smart, 1996.

5 Such influential authors include Ed Soja, Frederick Jameson and Mike Davis, Allen Scott and Michael Dear.

6 Box *et al* (1988) concluded that a number of factors are most significant in contributing to the fear of crime – vulnerability, environmental cues and conditions, personal knowledge of crime and victimisation, and confidence in the police. Fisher (1993) also proposed that methods to reduce the fear of crime could be seen in terms of risk avoidance, risk management and target hardening.

7 See Ellin (1997, back cover).

8 Cited in the *Electronic Telegraph*, 2 January 1996.

9 Cited in Oc and Tiesdell (1997).

10 Davies (1996b) indicated that discounts of up to 30 per cent have been obtained from insurers for CCTV installation.

11 Police Authority for Northern Ireland, opinion poll on CCTV (1995).

12 In Bentham's original idea, prison inmates were to be constantly under the gaze of the prison officer but could never see if they were actually being watched. An impression of omnipresence was constructed. This was intended to modify behaviour. It is commonly believed that the panopticon prison was never built. However, Wilson (1995) provides evidence from Cuba, from 1932, which shows such a design operating in practice.

13 Schmaltz (1988) cited in Jones and Lowrey, (1995, p.117).

Chapter 4

Risk Society and the Global Terrorist Threat

Introduction

Social scientists for many years have studied natural hazards and discussed the need to make contingency against their impact (White, 1942; Kates, 1962; Burton *et al*, 1978, 1993), whilst detailed study regarding the impact of technological risk has until recently not been undertaken to any great degree. Today, increasing emphasis is being placed on the analysis, assessment and response to both natural and technological risks, which threaten to alter the form and meaning of cultural landscapes and geographical relations on a variety of spatial scales (Blowers, 1997, 1998, 1999; Adam *et al*, 2000).

A number of accounts in the 1990s suggested that concerns about natural hazards and technological risks had become defining characteristics for organising contemporary society (Beck, 1992a, 1992b, 1997b, 1999; Smith, 1992; Douglas, 1994; Adams, 1995). Such accounts argued that 'risk' had an uneven distribution in both spatial and social terms and was increasingly creating definitive physical landscapes (such as flood defences or the proliferation of surveillance technologies in city centres) and social formations (such as environmental pressure groups or exclusive residential communities).

Over the last decade risk theory, has grown up primarily around concerns about global environmental hazards, and has been intimately linked to insurance, which for centuries has offered financial security against risk (Ewald, 1993; Adams, 1995). As starkly demonstrated by 9/11, it is only in recent years that the limits of insurance against certain risks are being pushed towards, and in some cases beyond, the limits of insurability. The severe environmental, social, economic and political implications of such risks are increasingly expressed through the media, and as such, the public are becoming increasingly aware of the trans-national nature of such risk. This has recently led to a series of academic and media debates surrounding notions of risk and their effects on social relations, the interaction between local and global processes, the collapse of the idea of the nation-state, and the rise of pressure groups that challenge the existing social and political order, redefining the rules and principles of decision making (Beck, 1996, 2002; Lianos and Douglas, 2000).

This chapter will first explore Ulrich Beck's *Risk Society* theory (in relation to the complementary work of Anthony Giddens) and consider how the nature of risk has changed in recent years and is affecting the relationship between society and space. The second part of the chapter deals specifically with one type of large-scale risk, namely the terrorist threat both during the 1990s and after 9/11. The relationship between aspects of risk society theory and the changing nature of the terrorist threat will then be discussed. This draws on a re-working of Beck's ideas relating to economic terrorist targeting prevalent in the 1990s (Coaffee, 1998, 2000a, 2000b) as well as Beck's own account of the changing state of *Risk Society* after 9/11 (Beck, 2002).

Contemporary Risk Theory

Until recently the social and cultural factors involved in discussions about risk have been hidden beneath a preference for an objective and rational approach to risk assessment (Ewald, 1993; Douglas 1994). Risk was seen as 'systematically caused, statistically describable and, in this sense, a "predictable" type of event, which can therefore also be subjected to supra-individual and political rules of recognition, compensation and avoidance' (Beck, 1992b, p.99). Today, risk has evolved into a concept that goes well beyond the idea of financial loss, although it is still common for insurers to view risk entirely in monetary terms (Dickson and Steele, 1995).

Although risk is translated into objective financial terms by the insurance industry, it should also be viewed as a cultural expression which includes individual and public perception of intangible loss. Douglas (1994) for example, argued that risk perceptions are related to a whole series of cultural factors reflecting a number of economic and political values. Risk is seen as an unacceptable danger that is often economically, socially or politically articulated. From this perspective, risk comes, for example, from the actions of those countries, corporations, groupings or individuals that are perceived as 'bad', 'dangerous' or simply as 'other'. This has particular resonances with how terrorist risk has been viewed post 9/11 with the demonisation of certain countries and regimes as 'the axis of evil'. For example Johnston (2002, p.221), citing examples from the rhetoric employed by US president George Bush, highlights that terrorists are commonly seen as 'evil doers' 'evil', 'evil people'... the 'evil ones', implying they are an enemy which must be eliminated.

In short, risk has become increasingly prevalent in today's society, because as we increase our knowledge about the cause and effect of particular risk events we become more aware that such events are inherently unpredictable and chaotic.

Risk Society

In the 1980s Ulrich Beck began to consider what society might look like when disputes and conflicts about new types of risk produced by industrial society are fully realised. In 1986 he published *Risikogellschaft* in German, which was subsequently translated into English as *Risk Society - Towards a New Modernity* (1992a). Beck's writing was influenced by a wide variety of social theory such as Max Weber's theories of bureaucracy, Habermas' ideas of late capitalism, Foucault's work on the power of institutions, Marx's ideas about the crisis of capitalism, Horkheimer and Adorno's 'Dialectic of Enlightenment', lesser known work by Ewald on the development of an 'assurance society', and even the Chicago School's project of 'social ecology' (Beck, 1995, p.41).

Beck's work has provided the impetus for academic research in a number of disciplines in the social and human sciences related to the impact of the emergence of a set of newly defined 'mega-scale' risks on the workings of western society. As Blowers (1999, p.256) noted, 'Risk Society is a pessimistic and conflictual diagnosis of modern societies… that is exposed to risks from high technology… that imperil our very survival'. The risks Beck referred to had a diverse nature and 'cannot be delimited spatially, temporally, or socially: they encompass nation-states, military alliances and all social classes, and by their very nature, present wholly new kinds of challenge to the institutions designed for their control' (Beck, 1995, p.1).

Beck provided a novel critique of contemporary risk. However, his work was criticised primarily for apparent vagueness and pessimism in terms of an irreversible process of degeneration. Leiss (1994), for example, saw Beck's work as over-emphasising risks in relation to the benefits of technological progress. Beck's work he noted had the 'overtones of an irrational zero-risk mentality' and was a diatribe against the rapid growth of risk management. Blowers (1999, p.257) also critically describes Beck's theory as 'a fatalistic acceptance of risks for which all are responsible but which individually we are unable to control', whilst Adams (1995) saw Risk Society as being 'no longer concerned with attaining something "good", but rather preventing the worst' (see also Carter, 1993; Hall, 1994; Boyd, 1995). Despite the criticisms, Beck's work does illuminate a number of issues, which have direct relevance to this book:

- The transition from modernity to a different type of society based on the perception of risk in an increasingly globalised society;
- The role of the media in the identification and social construction of risk and dangerous landscapes;
- The intense social criticism of the institutions of society;
- The spatial distribution risk and the creation of distinctive cultural landscapes;
- The processes by which new risks reflected by this new distribution pattern are denied the financial security of insurance coverage.

Each of these will now be discussed in turn.

Towards a New Modernity

The central theme of Beck's work was that society is in a period of 'transition' towards a 'New Modernity' in which the logic of industrial production and distribution based on wealth is becoming increasingly tied to the social production of risk. Beck (1996, p.27) argued that as industrial society has advanced, a Risk Society emerges which creates hazards of global magnitude. This he described as:

> A phase of development of modern society in which the social, political, ecological and individual risks created by the momentum of innovation increasingly elude the control and protective institutions of modernity.

Risk Society he argued has thus emerged from industrial society (which Beck terms a residual Risk Society) where security and safety have become increasingly important variables affecting the distribution of society, which previously, had been primarily influenced by economic factors. In industrial society, equally destructive risks were produced but were not subject to social criticism and politicisation. Beck saw this industrial phase as being dominated by societal 'self identity' and a phase of industrial production that produced 'residual risks'. In short Risk Society has emerged as the hazards produced by industrial society are increasingly criticised by the society that once legitimised and accepted them. As Beck (1996, p.27) argued, 'industrial society sees and criticises itself as a Risk Society'.

In *The Consequences of Modernity* (1990) and later in *Modernity and Self Identity* (1991), Giddens adopts aspects of Beck's Risk Society theory in which he too argues that modern societies are essentially risk societies. According to Giddens modern societies should be seen as double-edged, where, security versus danger and risk versus trust are key relationships, which are organised, differently in traditional and modern societies. Giddens argued that in traditional societies, localised trust relations existed based on an innate need for a stable social structure and safety within a territory. He proposed that in the modern world trust is increasingly forged from abstract, expert and institutional systems, meaning the control of the locale or territory is increasingly taken out of the hands of local people. In Giddens' words, 'ontological security' has been replaced by 'ontological insecurity' as local trust schemas are replaced by expert knowledge systems. Furthermore:

> Attitudes of trust, in relation to specific situations...are directly connected to the psychological *security* of individuals and groups. Trust and security, risk and danger: these exist in various historically unique conjunctions in conditions of modernity.
>
> (Giddens, 1991, p.19)

Giddens further argued that there is a danger of individuals and communities becoming 'disembedded' from the locale by globalisation, so that specific social circumstances of place become subsumed by global or national concerns for security. To give a sense of the foreboding that the modern world has introduced, Giddens (1990, p.139) employed a metaphor of an out-of-control juggernaut – 'a runaway engine of enormous power which collectively as human beings, we can drive to some extent but which also threatens to rush out of control'.

Both Beck and Giddens therefore talked of a new and radicalised modernity and referred to tendencies which enhance and expand modernity to a global scale but which have direct local affect. As Giddens further noted in *Runaway World* (1999, p.34):

> Whichever way you look at it we are caught up in risk management. With the spread of manufactured risk, governments can't pretend such management isn't their business. And they need to collaborate, since very few new-style risks have anything to do with the borders of nations.

Risk and the Media

Beck's thesis, although not explicitly devoted to the role of the mass media in the social evaluation of risk, does contain a number of 'buried references' (Cottle, 1998) which highlight how the media can be seen as important in the social construction, contestation and criticism of global risk. As Beck (1992a, pp.22-23) indicated:

> They [risks] can thus be magnified, dramatised or minimised within knowledge, and to that extent they are particularly open to social definition and construction. Hence the mass media and the scientific and legal professions in charge of defining risks become key social and political positions.

Furthermore, Beck (1995, p.142) continued by noting how the importance of the media affects 'profit opportunities' within a Risk Society:

> In a Risk Society...markets are built upon the card-houses of relations of definitions, and these can be knocked down merely by the wind in the mass media or by changes in public perception.

In this situation the media also provide the site for the contestation of risk. As a Risk Society develops, 'the social and economic importance of knowledge grows similarly, and with it the power over the media to structure knowledge and disseminate it. The Risk Society in this sense is also the science, media and information society. Thus new antagonisms grow up between those who produce risk definitions and those who consume them' (Beck, 1992a, p.46). This 'gives the mass media a leading role in sounding the social alarm' (Beck, 1995, p.100). He continues:

> The hazards, which are not merely projected onto the world stage, but really threaten, are illuminated under the mass media spotlight. The public want to be entertained; it pays, and delights in the thrills and spills of the real-life technological thriller, in which theatre and reality have changed places worldwide. In the grammar of entertainment, the superlative of the thriller is the daily news.
>
> (ibid. p.101)

As well as forming a key feature of the social construction and contestation of risk in a Risk Society, the media are also central to the criticism of technological risk; whilst at the same time inward investment and promotional agencies are attempting to actively construct a non-risk view of a particular society and space.

> The technocracy of hazard squirms in the thumbscrews of the safety guarantees which it is forced to impose on itself, and tightened time and time again in the mass media spotlight.
>
> (Beck, 1995, p.1)

The media in this sense perform a critical surveillance role for a Risk Society (Cottle, 1998), illuminating under the media spotlight hazards that are deemed threatening such as global warming, nuclear weapons and terrorist attack (Beck, 2002).

Reflexivity

Another key aspect of Beck's thesis concerns the concept of reflexivity – the ability of individuals and institutions to produce, through a consideration of the past and present, knowledge's about the future which will affect current practices. Burgess (1999, p.149) argued that reflexivity is now a key concept in the social sciences, suggesting that 'individuals and organisations constantly monitor their behaviour and experiences, and make adjustments in the light of new information'. Beck argues that through reflexivity, society can adapt to new risks. As Blowers (1999, p.257) noted, Beck, 'in this unrelentingly bleak portrayal of modern society… offers a small ray of hope in the form of reflexivity'.[1]

Beck argued that Risk Society is also associated with a new attitude towards scientific expertise where society increasingly has to place their trust in expert systems, which tell them what is safe or unsafe.[2] In relation to this, Beck often cited the work of François Ewald and the emergence of the 'assurance state'.

> The effective reality of a risk, that which "creates" the risk, is the contestation to which it may give rise... Some are accepted, others are not. Are some rejected because they are more serious, more dangerous than the others? Decidedly not. The idea of an objective measure of risk has no meaning here; everything depends on the shared values of the threatened group. They are what gives risk its effective existence.
>
> (Ewald, 1993, p.225)

Ewald continued by indicating how the general public has been socialised by such a system of shared values. The only conclusion is 'an acceptable risk is an accepted risk' (Ewald, 1993, p.285) As Smith (1992) further noted that the concept of acceptable risk is linked to the relationship between voluntary and involuntary risk taking, and that this leads to a certain risk tolerance being accepted. Using Ewald's ideas Beck (1995, p.92) highlighted that:

> All this serves to qualify the purely technological calculation and containment of risks, since the calculations are no longer thought of as arbitrators but as protagonists in the confrontation, which is enacted in terms of percentages, experimental results, projections, etc. Risks are social constructions disposing over technological representations and norms. An acceptable risk is, in the last analysis, an accepted risk. In the process, what appears unacceptable today may be routine tomorrow, while previously quotidian practices suddenly fill one with anxiety and terror in the light of new data.

Society therefore comes to terms with living with certain risks through reflexivity. This Beck (1995, p.94) noted, relates to the balancing of statistical measures of risk and cultural acceptance – 'without cultural standards, all calculations remains empty; without science and experimental results, any cultural stance is adopted blindly'.

However, the control by expert knowledge is questioned, said Beck, who highlighted disagreements between experts as to the assessment of risk, where threats and hazards have the ability to continually reshape public attitudes towards particular risks:

> Risk society is tendentially a self-critical society. Insurance experts contradict safety engineers. If the latter declare a zero risk, the former judge: non-insurable. Experts are relativised or dethroned by counter-experts... The former can be challenged by the latter, inspected, or even corrected.
>
> (Beck, 1996, p.32-3)

As such, experts are seen as attempting to shape maximum acceptable risk levels to allow business, commerce and ultimately globalisation to progress as the 'the destinies of markets, and hence companies… depend on them' (Beck, 1995, p.94).

Distributing Good and Bad Risks

In Beck's Risk Society, assessments of risk made by institutions and the media have helped create a society in which risk distribution is central. As Beck (1992a, p.12) indicated, 'in classical industrial society the "logic" of wealth production dominates the "logic" of risk production, in the Risk Society this relationship is reversed'. He noted that the concerns in industrial society with the distribution of wealth and useful resources (which in part has been eroded by the success of the welfare state in reducing scarcity) have been replaced by a quest for the avoidance of risk and uncertainty, and the need for safety creating distinctive new landscapes based on risk aversion. As Beck (2000, p.103) further noted, such threats can 'develop a society-changing power precisely in places where they have not appeared and put into action the underlying political meaning of risk dramaturgy, [not] to act before it is too late'.

Beck also argued, from the perspective of insurance, that the present drama of risk conflict in society is related to the confusing and contradictory ways in which 'new bads' interact with the old distribution based on 'goods'. In industrial society there was a strong emphasis on conflicts involved with the distribution of 'good' risks. In Beck's Risk Society, the distribution of 'bad' insurance risks are superimposed upon this pattern. Subsequently, institutions, most notably insurance, which are involved with the management of risk, have sought to find solutions to the new dangers. This, as Beck (1996, p.28) stated, poses 'questions of accountability' related to the way in which risks are 'distributed, averted, controlled and legitimated'.

In this situation, governments and institutions begin to lose their historical foundations and legitimacy. Their ability to manage contemporary risk is reduced and the need to privately employ risk management becomes increasing accepted by individuals and social groups, who then actively attempt to limit the affects of risk (a social 'bad') of their lives, for example, by moving to 'safer' areas. As Beck (1992b, p.112) noted:

> No matter how abstract the threats may be, their concretisations are ultimately just as irreversible and regionally identifiable. What is denied collects into 'loser regions', which have to pay the tab for the damage and its 'unaccountability' with their economic existence.

In contrast to Beck, Giddens (1994) provides a less pessimistic approach to risk distribution. In relation to Beck's concept of distributing 'goods' and 'bads', Giddens (1994) developed a concept which he called 'active trust' (which he has developed from work on ontological security) where new forms of social solidarity are developed through reflexivity and seek to deal with the occurrence of contemporary risk by either acceptance or affirmative action. For Giddens, 'modernity alters both the objective distribution and the experience of risk, and adaptive reactions to risk range from pragmatic acceptance to radical engagement' (Colomy, 1991, p.799).

When, relating Risk Society theory to crime prevention Ericson and Haggerty (1997) in their book *Policing the Risk Society* argued that the police operate in distinctly different ways in modern society given the sheer diversity of risks they have to deal with, and as such they are now increasingly forced into partnerships with other institutions that monitor risk. As Beck showed there are increased institutional demands for knowledge about risks in terms of definition, management and assessment, and as such the increasing involvement of welfare agencies, health authorities, the risk management profession and most notably insurance companies actively changes the way in which risk is defined and managed by the police. Ericson and Heggerty (1997, p.6) further noted Beck's work on the distribution of good and bad risks, showing how society increasingly focuses on the fear of 'bads' rather than on the progress of 'goods'. They suggested that as fear within society increases it drives us to seek ever more intricate ways of judging risk. As Hope and Sparks (2000, p.2) further note, Beck's Risk Society thesis 'impute to late modern citizens an array of concerns and worries that suggest a permanently unfulfilled quest for security'.

They further argue that the police are no longer seen as the sole agency of social control but as part of a widening fragmentary web of surveillance and control, which attempts to reduce the risk of 'bads'. They also note the heightened importance placed on community based law enforcement where through 'communications policing' the police can encourage residents and businesses to solve their own risk and security problems or employ private security (Jones and Newburn, 1998).

Beyond the Insurance Limit

As noted in the previous section, the distribution and control of risk is intimately linked to the provision of insurance within a Risk Society. Risks, argued Beck (1994, p.181):

> Are an attempt to make the incalculable calculable. Events that have not yet occurred become calculable (at least economically) through the insurance principle.

Risk Society is viewed as the return of uncertainty, where problems of order are now perceived as problems of risk to which there is no certain solution. In short, one of Beck's key arguments is that in a Risk Society it is not possible to insure against all types of risk. Beck (1995, p.107), citing Ewald (1986), argued how the production of security within a society is a sociological construct based on how institutions deal with the dangers of Risk Society:

> From this perspective, risk calculations and private and state insurance policies are social answers to the challenge of the insecurities created by modernity in every area of life.

Social progress has led to the establishment of diverse systems of insurance to provide a 'social pact' (Beck, 1995, p.108) to counteract insecurity, where society as a whole can be seen as a 'provident state':[3]

> The institutions of developing industrial society can and must also be understood from the point of view of how the self produced consequences can be made socially calculable and accountable and their conflicts made controllable. The unpredictable is turned into something predictable... The dialectic of risk and insurance calculation provides the cognitive and institutional apparatus.
>
> (Beck, 1996, pp.30-31)

Contingency planning by society thus comes in the form of an insurance contract, which objectively assesses the potential risk using statistics to put a financial value on projected loss. As Beck (1995, p.109) cited:

> In this way the 'assurance state' arises as a pendant to 'Risk Society'; a social architecture whose social integration, idea of justice and social contract become sociotechnically shapeable and perfectible...according to the model of risk and assurance.

Beck attempted to indicate the historical moment when such risk assessments become commonplace. This, he noted, occurs when industrial society becomes a Risk Society, and when a fully insured society becomes impossible as some risks become incalculable to the insurance industry. As Beck (1996, p.31) indicated, 'the entry into Risk Society occurs at the moment when the hazards which are now decided and consequently produced by society undermine and/or cancel the established safety systems of the provident states existing risk calculations'. As such:

> Industrial society, which has involuntarily mutated into Risk Society through its own systematically produced hazards, balances beyond the insurance limit.
>
> (ibid.)

The residual risk (industrial) society has thus become an uninsured society, with protection paradoxically diminishing as the danger grows (Beck, 1992b). Within such a society it is the insurers in particular who judge the limits of Risk Society:

> It is the private insurance companies which operate or mark the frontier barrier of Risk Society.
>
> (Beck, 1996, p.31)

To clarify this, Beck (1992b, p.103) drew a further distinction between actual risks and threats within such a society relating this to insurance:

> Is there an operational criterion for distinguishing between risks and threats? The economy itself reveals the boundary line of what is tolerable with economic precision, through the refusal of private insurance. Where the logic of private insurance disengages, where the economic risks of insurance appear too large or too unpredictable for insurance concerns the boundary that separates 'predictable' risks from uncontrollable threats has obviously been breached again and again in large and small ways.

From a more institutional perspective, Giddens (1991, p.29) also showed how the strategy of objective statistical risk assessment and an attempt to predict or 'colonise the future' has been built into contemporary institutions:

> Insurance, for example, has from early on been linked not only to the risks involved in capitalist markets, but to the potential futures of a wide range of individual and collective attributes. Futures calculations on the part of insurance companies is itself a risky endeavour, but it is possible to limit some key aspects of risk in most practical contexts of action...and such companies typically attempt to exclude aspects or forms of risk which do not conform to the calculation of large-sample probabilities.

Giddens here hinted strongly at the practice of 'redlining', where the insurance industry and other financial services discriminate against certain risks and certain high-risk geographical areas. Such insurance practice will be highlighted in Chapter 6 in relation to terrorism insurance coverage, where insurance is seen as a mechanism for distributing risk. As Giddens (1999, p.25) explained:

> Insurance is the baseline against which people are prepared to take risk. It is the basis of security where fate has been ousted by an active engagement with the future… Insurance is about providing security, but it is actually parasitic upon risk and people's attitudes towards it… Those who are providing insurance… are essentially simply redistributing risk.

Insurance and the Urban Landscape

In recent years a number of specific urban risks have generated much concern for the insurance industry, leading to the exclusion of certain risks and certain geographical areas from policy protection. Urban risk in this sense encapsulates concerns over flooding, subsidence and atmospheric pollution, as well as socio-cultural risks such as fear of crime, rioting and in extreme cases terrorist attack. Within this context, Risk Society theory can be applied at a city-wide level in order to study the affects of the occurrence and distribution of risk. The occurrence of such risk can also illuminate how the actions of particular institutions (most notably insurance) can be seen as key urban managers seeking to influence the relationship between society and space.

Insurance is one possible method of reducing financial uncertainty and risk transference (Dickson and Steele, 1995). As Diacon and Carter (1995, p.7) indicated, the relationship between risk and insurance is 'one where you exchange a situation of risk (where different financial outcomes are possible) for one of financial certainty (that is, with only one definitive financial result), since the insurance company guarantees the purchaser, subject to certain provisos, that his financial position will not be affected by the occurrence – or non-occurrence – of certain specified events'.

There are certain economic factors which determine whether or not a particular risk is deemed insurable. First, the insured risk must be measurable in monetary terms to allow an accurate premium to be charged. Second, the exposure to a particular risk must be what insurers call homogeneous; there must be a sufficiently large number of separate and independent exposures to allow the insurance companies to compile objective statistics for a particular risk. In the event of a risk incident occurring this allows the cost to a few to be covered by the premiums of many. This will allow the insurer to remain solvent. Premiums obtained must therefore be from both high and low risk areas. In short, insurers attempt to ensure that the properties they cover are geographically spread so that only a small part of the total exposure can be damaged or destroyed in a single event.[4]

Insurance therefore works by transferring the risk of loss from one person, or area, to another. Insurance is a mechanism for sharing risk – in essence, everybody pays a little so nobody is forced to pay much. However, not everyone pays the same. Individuals are expected to pay premiums in relation to the level of risk each is perceived to represent. Such premiums can be reduced as the insurance industry can also directly influence the risk of loss by encouraging clients to install loss prevention and risk management measures for which the insured often receive a reduced premium.

During the last 30 years in Britain the insurance industry has become ever more significant in shaping the physical landscape of the city. In the early 1980s, the losses imposed by inner city riots led the insurance industry to make stipulations about building design and security measures that materially influenced the shape and meaning of the regenerated urban landscape (Patel and Hamnett, 1987). Perhaps more significant was the fact that domestic and commercial insurance cover also became difficult to obtain, which affected the flow of investment into the areas most affected by the riots.

Today, there are certain risks, as well as certain areas of cities, where insurance companies will not offer an affordable premium due to a unfavourable ratio of claims to premiums. The insurance industry can thus be seen to discriminate against 'bad risks' on the basis of economic rationality, a practice which has been termed 'redlining', and which received substantial attention in North America during the 1970s and 1980s (see for example Advisory Committee to the NAIC Redlining Task Force, 1978; American Insurance Association, 1978; Squires and Valez, 1987; Squires *et al*, 1979) and more recently in the UK.[5] Despite the origins of insurance redlining being related to specific geographical areas, it can now also be seen to relate to specific risks, which have trans-national impacts but which are still linked to certain localities.

UK research has often been critical of redlining practices arguing that negative stereotyping of inner-urban areas unjustly penalises the businesses and residents located there (Feildstein, 1994; Threadgold, 1995, 1996). Critics believe that, both overtly and covertly, insurance policies can be seen to benefit certain social groups at the detriment of others. It has been argued that such discrimination can have a key role in determining the viability of urban communities, as it constitutes a major element reinforcing, promoting and contributing to uneven urban development and thus the continuing restructuring of the city into a patchwork of safe and unsafe territories (Leyshon and Thrift, 1995; see also Leyshon and Thrift, 1994, 1997).

Critics of redlining also point out that by selectively 'cherry-picking' their risks the insurance industry is operating with rating classifications that are too small to be statistically valid. The insurers have also been criticised for generalisations they make about particular neighbourhoods or post/zip-code regions, which deem all properties uninsurable by association.[6]

Furthermore, it has been argued that subjective perceptions of city areas, and of certain social groups within these areas, make the insurance industry an important 'shaper' of the social conditions that prevail in the city. A number of other studies have supported this assertion showing such financial exclusion to be an important factor in the development of the city. The inability of inner-city residents and businesses to obtain insurance was initially highlighted after the inner-city riots of the early 1980s and the mid-1990s (Threadgold, 1995). In short 'the unfortunates who live in the high crime zone are consigned to the insurance scrapheap. Many insurers have been unwilling to insure them at any cost' (Scott, 1994).[7] For example, using leaked material from London insurers, the Association of London Authorities (ALA, 1994) accused the insurance industry of redlining in certain London districts considered to be high crime, high risk or high claim areas (ALA, 1994). This report also indicated that redlining is a continuing problem in many major cities in the UK as a result of further civil unrest. For example, the 1995 urban riots in Bradford and Brixton again focused attention on the insurance industry, who indicated that negative media stereotypes of the inner city were to blame for criticism aimed at them. As a leading insurer highlighted, 'we underwrite business in every postcode in the country. We don't redline anywhere including Moss Side, Manchester, Toxteth or Liverpool. Toxteth isn't as bad as you would think from the media' (cited in Threadgold, 1995). Furthermore, using the example of ground subsidence in London, Doornkamp (1995) showed how the perception of the insurance industry is related to experience and elapsed time since the last catastrophic event. In the UK the decision of the insurer to offer coverage appears to depend on a subjective or institutional perception of the dynamics involved and not necessarily the reality of the situation, that is the statistical likelihood of a risk event occurring.

Redlining and Risk Society

Whereas the original ideas behind the insurance redlining of home contents and business insurance are now basically understood by urban commentators, a new set of risks which have important geographical consequences have come to dominate the insurance and reinsurance agenda during the 1990s and the early 2000s. These catastrophic risks include both natural hazards such as hurricanes and earthquakes as well as technological risks such as chemical and nuclear leakage, the effects of global warming, and large terrorist attacks. Assessment of such risks shows that insurers and reinsurers, anxious to protect their solvency and profits, now see redlining as a strategy that can be increasingly used on both a local and global scale to cope with new types of risk. Today's insurers, faced with redlining accusations, operate a policy of 'adverse selection' attempting to limit their liability to mega-scale risks. This is increasingly done through association with a national government who bears part of the risk. Earthquake cover in Japan (Morimiya, 1985), and general disaster protection in New Zealand (Smith, 1992) are examples of this.

Insurers are no longer able to offer policies on certain risks that break the fundamental rules of insurance, as they are immeasurable financially, non-homogeneous and not limited in time and space. As such, in Beck's terms, we have entered into a Risk Society. These ideas will be explored in Part II of this book in relation to the provision of terrorism insurance cover in the UK.

Revisiting Risk Society Post 9/11

Although Beck's conceptual work in the 1990s covers many aspects which were applied to the impact of large-scale terrorist risk (Coaffee, 2000a, 2000b, 2002), Beck himself did not make any substantive comment upon this issue until after 9/11. In *The Terrorist Threat: Risk Society Revisited* (Beck, 2002), Beck notes that global terror networks and the way in which they empower governments and national states has become a new axis of the world Risk Society since 9/11. In particular he argues that notions of 'trust are replaced with mistrust and as such "the terrorist threat" triggers a self-multiplication of risks by the de-bounding of risk perceptions and fantasies' (ibid. p.44). He argues that the key question is 'who defines the trans-national terrorists?' because this of course will ultimately determine outcomes and possible reprisals. Beck argues that such enemy images are a gross simplification and are constructed by security services and government departments without significant public discourse. This replaced a far more open system of decision making about risk which characterised former large scale risk where arguments between experts was common place and public debate forthcoming (ibid.).

Beck also draws attention to what he calls the 'speed of acknowledgment'. Whereas certain global environmental risks are not recognised or at least disputed (for example the rate of global warming), terrorist risk has a far greater immediacy:

> With the horrific images of New York and Washington, terrorist groups instantly establish themselves as new global players competing with nations, the economy and civil society in the eyes of the world. The terrorist threat, of course, is reproduced in the global media.
>
> (Beck, 2002, p.45)

In short, when dealing with the actual dynamics of Risk Society, Beck offers the following argument in relation to terrorism:

> To summarize the specific characteristics if terrorist threat: (bad) intention replaces accident, active trust becomes active mistrust, the context of individual risk is replaces by the context of systematic risks, private insurance is (partly) replaced by state insurance, the power of definitions of experts is replaced by threat of states and intelligence agencies; and the pluralisation of expert rationalities has turned into the simplification of enemy images.
>
> (Beck, 2002, p.45)

The Global Terrorist Risk

Large-scale risks, such as those Beck refers to, are increasingly denied insurance coverage and become subject to high levels of exposure in the media. These risks necessitate that society, and in particular its key institutions, act reflexively to determine whether or not the risk is acceptable, and hence accepted, or whether risk management measures should be implemented in an attempt to reduce the occurrence, and fear, of particular risks. Such risk-reducing measures are most commonly undertaken at the sites of greatest risk, or 'loser regions' as Beck terms them, often serving to alter the material landscape significantly. This section of the chapter will highlight how terrorism against specified economic targets fits into this Risk Society scenario.

Defining Terrorism

In the last thirty years, terrorism, given the growth of the mass media, has become an issue of worldwide attention and a subject on intense analysis by politicians, security agencies and academics alike. This is especially the case in the aftermath of 9/11 where the academic, policy and popular press have been inundated with writings on terrorist related issues. However, the term terrorism is inherently subjective relating to the motivations of the perpetrators as well as the positionality and values of the commentator and viewer, and hence, definitions vary between, and within, cultures (Merari, 1993). Generalisations about what the concept of terrorism actually is should be treated with a good degree of scepticism not only because of the specific context in which each supposed terrorist act occurs but, importantly if we over-simplify the terrorist phenomenon there is a danger we also over simplify the counter-response to such acts. As Davidson-Smith (1990) noted:

> An accurate understanding of terrorism is obtainable through precise assessment in a given context. The assessment however must include the complexities of motivation,

organisation, methodology and desired goals. It is through a serious appraisal of these factors that the threat may be better understood and more effectively countered.

In attempting to clarify how we should regard terrorism, Schmid (1992, p.7) outlined four ways in which terrorism could be defined in which he attempted to escape what he called the defeatist position that 'one man's terrorist is another man's freedom fighter'. As such he distinguished between the academic discourse, the governmental position, the public perception, and the terrorist's own view. Other commentators on terrorist issues have, by contrast, used a variety of different classifications to characterise terrorist motivations, including 'cultural', 'ideological', 'criminal', 'nuclear', 'chemical', 'computer', 'moral' and most notably 'political', 'religious', and 'state-sponsored'. Although a detailed description of terrorism *per se* lies outside the scope of this book (for more information see Dobbs, 1990; Schmid, 1992, Hoffman, 1998) most definitions contain what Merari (1993) called the three 'cornerstones' of terrorism: violence, political motivation and installation of fear into the target population.[8]

Contested Meanings of Terrorism

As indicated above, the paradigm shift within the social sciences towards the acceptance of contested meaning has meant that defining terrorism depends on a subjective point of view (Chalk, 1996). Terrorism as a concept is multifaceted, although, in Western societies, certain powerful discourse communities dominate. In recent years this perspective has been taken forward by the work of anthropologists, Zulaika and Douglass (1996) in *Terror and Taboo: the follies, fables, and faces of terrorism*. The approach they adopt contrasts with that of most counter-terrorist experts or politicians who, they argue, view terrorism in terms of statistics and adopt a normative approach:

> The discourse of the terrorism expert is buttressed by the scientistic idea that true knowledge must afford the objectivity that allows one to talk about society in terms of universal criteria.
>
> (ibid. 1996, p.181)

They argued that how terrorism is viewed is contingent upon the socio-cultural, political and context in which it occurs, and, upon the viewer:

> We view terror as a shifting representation that commands diverse perceptions from different actors and audiences in separate situations. What is happening is simultaneously a *struggle* for supporters of the violence, *crime* for its detractors, *error* for those who know the actors too well, *stupidity* for those maintaining satirical distance.
>
> (ibid. 1996, p.89)

However, they argued that in Western countries, portrayals of terrorism are one-dimensional, which 'taboos' the concept and its alleged perpetrators, highlighting them as evil, deranged or religious fanatics. This they argue is related to 'academic fashioning, media consumption and political manipulation of the terrorism discourse' (ibid. p.xi). They further noted that:

> Some of the world's most powerful vested interests drive terrorism discourse for their own purposes – including the media in search of stories, academics enhancing career paths, filmmakers and novelists in search of plots, a multibillion-dollar security industry selling its services, and above all, a plethora of government agencies defending turf and budget.
>
> (Douglass and Zulaika 1998, p.265)

They highlighted that such discourses drive the need for counter-terrorism within society to deal with what is portrayed as an ever-increasing risk:

> Once something that is called "terrorism" – no matter how loosely defined – becomes established in the public mind, "counterterrorism" is seemingly the only prudent course of action. Indeed, at present there is a veritable counterterrorism industry that encompasses the media, the arts, academia, and, to be sure, the policy makers of most of the world's governments. There is now in fact an "official" line acknowledging that terrorism poses a global threat to world security, which in turn justifies the expenditure of billions of dollars on counterterrorism measures.
>
> (Douglass and Zulaika, 1996, p.ix)

Douglass and Zulaika (1996) argued that terrorism is articulated within society as an 'expert view' backed up volumes of statistical data. They noted however, that this is a biased process, as the figures given on different databases, controlled by different countries and organisations vary considerably as a result of differences in the way in which acts of violence are defined and categorised. They concluded by noting that 'statistical manipulation is therefore unavoidable. Yet such statistics are the backbone of the entire discourse' (ibid. p.23).

However we choose to define terrorism it is clear that it has become a major political weapon and has led to reactive and proactive and pre-emptive measures by the governments and organisations that are targeted. This has increased dramatically since 9/11 as buildings have been fortified and prominent individuals guarded at great cost to the taxpayer. National and international law has attempted to co-ordinate global action against international terrorist organisations and substantial use has been made of new technologies, especially in relation to intelligence gathering and surveillance. Many of these developments have subsequently caused concern amongst civil liberty groups who see the response of the authorities as an invasion of privacy (Gearty, 1991; White, 1991; Liberty, 2002).

The Media Role

As indicated in the first part of this chapter, the media is a central construct for defining risk. The media's strong interest in terrorism has meant that terrorists have often changed tactics to ensure the fullest media coverage. Conversely, the media can have a significant role to play in shaping public opinion against the terrorist by being supportive of the official government policy and the reactive military response (Alexander and Latter, 1990; Dobkin, 1992; Paletz and Schmid, 1992; Schaffert

1992; Picard, 1994; Nacos, 2002). As Wilkinson (1997, p.53) indicated, 'as long as the mass media exists, terrorists will hunger for what former British Prime Minister, Margaret Thatcher, called "the oxygen of publicity" and for as long as terrorists commit acts of violence the mass media will continue to scramble to cover them in order to satisfy the desire of their audiences for dramatic stories in which there is inevitably huge public curiosity about both victimisers and their victims'.

If terrorism is designed to have psychological effects and aimed to impact upon an audience, then public fear of terrorism is strongly influenced by the media. As Gearty (1992, p.9) further observed in relation to terrorism:

> The opportunity for communication with the wider audience... is, of course, one of the main reasons why it [terrorism] occurs in the first place. Society wonders who will be next and, in its weakened state is more susceptible to the political message of the moment... in this way terrorism springboards issues into public debate. It uses horror and fear to jump the queue of ideas waiting for public attention.

Media publicity is a key factor driving terrorist tactics, and whilst the media do not actively create terrorism, their actions 'may facilitate the strategic success of terrorist groups' (Alexander and Latter, 1990, p.3). For example, Hermon (1990, p.38) argued that in Northern Ireland Sinn Fein[9] and the Provisional IRA were 'waging two wars', these being the continuation of violence and the propaganda war through the media which are inseparable.[10] Furthermore, as Hoffman (1998, p.142) noted, 'terrorism and the media are bound together in an inherently symbiotic relationship, each feeding off and exploiting the other for their own purposes'. In summary, Wilkinson (1997) highlighted that the terrorist is trying to achieve four main objectives from increased media coverage: first, to convey the propaganda of the deed and to create extreme fear among their target group(s); second, to mobilise wider support for their cause among the general population and international opinion, by emphasising such themes as the righteousness of their cause and the inevitability of their victory; third, to frustrate and disrupt the response of the government and security forces, for example by suggesting that all their practical anti-terrorist measures are inherently tyrannical and counterproductive; and fourth, to mobilise, incite and boost their constituency of actual and potential supporters and in so doing to increase recruitment, raise more funds and inspire further attacks.

This chapter will now turn its attention to the actions of the Provisional IRA over the last thirty years, which provided the context for the defensive strategies deployed to protect the City of London.

Economic Terrorist Targeting in the 1990s

In a Risk Society Beck argued that the distribution of wealth is now being juxtaposed with the distribution of risk. As such, economic terrorist targeting provides an illustration of this with financial areas being bombed or fearful of attack. This was especially the case in the early 1990s where there was a perception that acts of economic terrorism were occurring more frequently in England, as well as globally (Rogers, 1996).

In the 1990s terrorist targeting was increasingly centred towards certain geographical targets related directly to key economic areas. This type of targeting was most prevalent in the early-mid 1990s with such attacks becoming less widespread in recent years (Oakley, 1995; Wilkinson, 1996; Johnson, 1997; Leader, 1997). However, at the beginning of the 1990s there was a growing realisation by the world's terrorists that by targeting business centres and their commercial infrastructure they could not only cause severe damage directly to valuable building structures, but also to the reputation of the area through media exposure. As Timothy Hillier of the City of London Police (1994) stated:

> Massive explosions in London, New York and other major cities world-wide clearly demonstrate that important financial districts have become prestigious targets for terrorist organisations, regardless of their motives. In addition to causing significant loss of life, these bombs severely disrupt trade and economic transactions. Further, modern satellite communications broadcast grisly bomb scene images around the world within minutes adding to the lure of this type of target for groups seeking media publicity.

Other prime examples of this included the World Trade Center bombing in New York in 1993 when a van bomb parked in an underground car park at the World Trade Center in Manhattan, New York exploded killing six, injuring thousands and causing extensive damage; the bombing of Central Bombay in 1993 when a series of 13 bombs were detonated in India's financial centre, Bombay, killing over 250 people; and the Tokyo subway attack in March 1995 when the Aum Shinri Kyo religious sect attacked the Tokyo subway system with improvised chemical weapons containing the nerve agent sarin (see for example Brackett, 1996).

As a direct result of the occurrence and fear of such attack, individual buildings as well as discrete commercial districts increasingly attempted to 'design out terrorism'. Patricia Leigh Brown (1995), writing in the *New York Times*, exemplified how defensive landscapes, based on Oscar Newman's defensible space principles (see Chapter 3) were constructed in certain parts of the city to prevent further acts of terrorism. She relates the responses of the New York Authorities after the 1993 World Trade Center bomb with ideas of controlling the urban area through ideas of defensive space, indicating how the threat of terrorism is becoming a key determinant of architectural form and urban design:

> After the World Trade Center was bombed in 1993, the principles of defensible space design were put into place there. In addition to concrete planters parking is no longer open to anyone. Tenant parking is controlled and includes a hydraulic barrier – a latter day drawbridge – lowered by a guard only after the proper credentials are shown and capable of stopping a truck at 50 miles an hour.

The responses of urban authorities and the agencies of security, is however, directly related to local circumstances of place and the tactics of the threatening terrorist group. As Hoffman (1998, p.205) notes, 'terrorism is among the most fluid and dynamic of political phenomena... constantly evolving into new and ever more dangerous forms in order to evade security procedures and surmount defence barriers placed in its path'. He continued by noting that effective counter-terrorism must also move with the times:

> Any government's ability to craft an effective response to terrorist attack…will inevitably depend on its ability to understand the fundamental changes that distinguish today's terrorists from their predecessors. Only in this way can the array of required counter-measures be first identified and then brought to bear with genuinely positive results.
>
> (ibid. p.206)

Targeting the British Economy

As noted in Chapter 2, during the 1970s, and to a lesser extent in the 1980s and 1990s, central Belfast was attacked by the Provisional IRA. Such attacks were specifically directed against economic targets and can be seen as a precursor to the 1980s and 1990s Provisional IRA bombings in London. Terrorist targeting in Northern Ireland during the 1970s aimed to disrupt the economy and took a variety of forms with attacks made against central business districts, energy and raw material resources, communication facilities, and transport infrastructure (Murray, 1982). Such attacks led to security measures being built into the physical landscape of Belfast and other towns.

During this time the commercial insurers also decided to withdraw coverage for terrorism and terrorist related risk in Northern Ireland as their financial liability was becoming too high. This created financial insecurity, and eventually forced the British government to pay all insurance and compensation claims. This, according to Beck, is indicative of a Risk Society where the insurance industry judges that a fully insured society is impossible in the given circumstances, necessitating risk management measures.

Although Belfast continued to be attacked during the 1980s the frequency of bombing reduced considerably. This was due to two key factors. First, there was a change in Provisional IRA tactics, which increasingly saw England as the key target and second, the belief by some that the success of the security cordon around Belfast city centre forced a change in Provisional IRA targeting priorities towards softer targets.

The Provisional IRA in London

The experience of Belfast provides a historical context for the economic targeting and resultant defensive landscape changes introduced in the City of London as a result of Provisional IRA activity in the 1990s.

However, it should be noted that economic targeting in England, and especially London, by the Provisional IRA was not a new phenomenon. The Provisional IRA has periodically targeted London since the 1930s, although the City itself was not specifically targeted until recently. As the writer Martin Dillon (1996, p.18) commented:

> For those of us that have watched the horror of the London bombings in the 1970s, 1980s, and 1990s it is difficult to believe similar scenes were part of the life of the city in 1939.[11]

It became clear to the Provisional IRA in the 1970s that a protracted bombing campaign in Northern Ireland would not put sufficient pressure on the British Government to withdraw from Ulster. They thus decided to extend their campaign to England in the hope of thrusting the 'Irish question' back into the centre of the political agenda. This would also have the affect of restricting bombings in Northern Ireland. As persons present at the June 1972 meeting of the Provisional IRA Army Council indicated, 'sooner or later there would have to be a drift to another area to take the heat off Belfast and Derry' (Bishop and Mallie, 1987, p.250).

According to Dillon (1996) the Provisional IRA Army Council[12] decided (in June 1972) that an English bombing campaign should be restricted to targets in central London with minimum civilian casualties. It is important to note that the Provisional IRA were not the only terrorist group attacking London at this time. In particular the left wing anarchist group the Angry Brigade were behind a series of explosions outside state buildings and at the homes of leading politicians and businessmen between 1968 and the end of 1971. This group were in large part responsible for generating fear amongst Londoners about the threat of terrorism (Davidson-Smith, 1990; White, 1991) and set the scene for the fear and disruption caused by Provisional IRA attacks.

The Provisional IRA campaign was finally given the go-ahead in early 1973. In 1973-4 there were a variety of attacks against different types of targets including attempted political assassinations. During this time England as a whole suffered over 100 Provisional IRA bombings, killing nearly 50 people. The aim of the campaign was to 'strike at the economic, military, political and judicial targets'.[13] This campaign was followed by a cease-fire between December 1974 and mid 1975. The summer of 1975 saw the renewal of a London bombing campaign with hotels and banks in central London being especially targeted. When this wave of bombings subsided, London was only targeted sporadically as it became evident that the bombings were not having the desired affect on the English population and indeed could well be seen as counter-productive, serving to alienate opinion against the Provisional IRA, as well as the Irish population in England, particularly in London (Bishop and Mallie, 1987).

The bombing of the Grand Hotel in Brighton in 1984 during the Conservative political party conference effectively brought to an end another bombing campaign (1981-1984) aimed at key mainland targets. The arrests of Provisional IRA suspects following this event broke up the terrorist cells (Active Service Units, or ASUs) acting in and around London and as a result no more attacks occurred in 1985-6. The creation of new ASUs in 1987 led to a number of high profile bombs in London between 1987 and 1990. It is at this stage that military and political targets were singled out such as attacks on Army career centres and the assassination of prominent MPs (Clutterbuck, 1990).

In the late 1980s there appeared to be a shift in Provisional IRA tactics towards attacking non-civilian targets. The British security services see the bombing of the

Mill Hill army Barracks, in North London in August 1988 as the moment when the Provisionals began to move away from previous strategies. By the early 1990s the majority of bomb attacks were against the industrial, commercial or transport infrastructure – in short, economic targets. It could be argued that a more effective strategy would have been to specifically target just transport and communication infrastructure, as this would create mass disruption with minimal chance of being caught. However, economic targets were increasingly favoured for the media attention such attacks received. As Dillon (1996, p.265) commented:

> Political assassination was always favoured by the IRA, but their main aim for the 1990s was to bring terror to the heart of London with a ferocity never before experienced in the capital.

As Rogers (1996, p.15) further commented:

> In the early 1990s PIRA continued with a range of paramilitary actions… but there was a progressive move away from the deliberate targeting of civilians and towards economic targeting.

From a British perspective the Provisional IRA successfully attacked a number of key economic targets in London in the 1990s. Large bombs exploded in the City in April 1992 and April 1993. In November 1992 a bomb was found under the Canary Wharf Tower in the London Docklands, and in February 1996 they succeeded in bombing South Quay Station in the southern part of the London Docklands. These bombings and the subsequent reaction of urban authorities, the police and the insurance industry provide the context for the remainder of this book.

Conclusions

The nature of potential terrorist threats is rapidly changing and with it must change the responses made to counter it. Contemporary terrorism now encompasses the potential use of weapons of mass destruction, specific types of criminal activity, hostage-taking, info-terrorism, political assassination, suicide attacks and even civil rights protest. New terrorist realities has subsequently led to a change in thinking by policy makers and security professionals with new working definitions and new ways of thinking, seeing and responding to terrorism being developed as a response to the catastrophic potential of terrorist attack.

In 1996, Walter Laqueur published a small article in *Foreign Affairs* entitled *Postmodern Terrorism.* At the time of publication the article, although well received, was one of many which highlighted the potential dangers of weapons of mass destruction (WMD), especially those in the hands of 'fundamentalist' terrorists. Until 1995 such fundamentalist groups were not taken that seriously by the authorities. All this changed after the Aum Shinro Kyo (Supreme Truth Sect) poison gas attack on the Tokyo subway system, which indicated that such groups could relatively easily influence the workings of government and spread terror in society (see for example Juergensmeyer, 1997). Such groups were then increasingly seen as a potential high-risk threat by the security services in Britain. Indeed soon after this attack it is reported that known members of the Provisional IRA were seen in Tokyo, it is suspected, planning a similar attack on the London underground. Apart from the Tokyo attacks there was a lack of actual terrorist incidents involving what could be described as WMD, and subsequently the ideas put forward by Laqueur were mainly put to one side.

Not unsurprisingly, since 9/11 when the truly international nature of the terrorist threat came to the fore, Laqueur's paper has been the source of much discussion with the notion that there were indeed 'new rules for an old game' (p.24). In particular the notion of 'asymmetric threats' dominates the contemporary terrorism discourse, where 'wars are increasingly fought between hugely unequal powers but with the apparently week able to inflict massive blows on the apparently powerful' (Urry, 2001, p.61).

According to Laqueur (1996) acts of 'terrorism' appear to be carried out by an ever-growing number of 'aggressive movements espousing varieties of nationalism, religious fundamentalism, fascism and apocalyptic millenarianism...' (ibid. p.28), with different motivations, aims and approaches. The methods adopted by these 'postmodern groups' differ in a number of ways from more traditional conceptions of terrorist groups. For example, the aim is often to cause mass casualties which can be justified through particular readings of religious texts. Such groups also operate outside of the traditional nation-state model given the porous nature of state boundaries and the widespread use of ICT, especially the Internet, which allow groups to operate from transient bases. This sets up a scenario where 'terrorism becomes the dark side of globalisation' (Colin Powell, 2001, cited in Urry, 2002). Perhaps, prophetically, in the light of 9/11, Laqueur (p.36) further noted:

> Of 100 attempts at terrorist superviolence, 99 would fail. But the single successful one may claim more victims, do more material damage, and unlash far greater panic than anything the world has yet experienced.

The unprecedented physical, financial and psychological damage of 9/11 will, many commentators have hypothesised, have a lasting impact on the way society views terrorism and how our major cities are planned, run and function (see for example Savitch and Ardashev, 2001; Grover, 2002; Marcuse, 2002a, 2002b). As Graham (2002) notes 9/11 'has underlined once again, the critical roles of cities as key strategic sites of military, economic, cultural and representational struggle as we enter this quintessentially urban century' (p.589).

As such commentators have posed questions about whether we should rethink urban development strategies on the basis of 'worst case scenario' terrorism. Should we seek to generate a 'bunker mentality', construct an 'architecture of fear', create 'exclusion zones', '*cordon sanitaires*' or modern-day 'panopticons', on the basis of what might, or might not, happen? Will such security schemes if developed be acceptable to the public? It is interesting to note that in the aftermath of 9/11 eighty percent of Americans questioned in a *New York Times*/CBS poll indicated they were prepared to have less personal freedom if it meant the country as a whole could be made more secure further terrorist attack (cited in Rosen, 2001).

The subsequent chapters will highlight how key urban managers in the City of London have sought to 'design out terrorism' over the previous decade, and how this process was contingent upon local histories, geographies and institutional arrangements as well as the influence of the global economy. They will also highlight how the defensive measures employed were constantly altered in relation to the perceived threat level and changing tactics of the Provisional IRA and other terrorist threats and how the provision of insurance coverage against terrorist attack influenced the physical landscape changes that were subsequently developed.

Notes

1 When dealing with the emergence of risk society Beck utilised the concept of 'reflexive modernisation' (see Beck, 1992a; Beck *et al*, 1994), where: 'Modernisation within the horizon of *pre*-modernity is being displaced by *reflexive* modernisation... We are witnessing not the end but the *beginning* of modernity – that is, of a modernity *beyond* its classical industrial design' (Beck, 1992a, p.10).
2 In Beck's risk society, science is seen as the controller of technology, which cannot ultimately be controlled, as we are only just beginning to fully realise the hazards associated with modernisation. However through calculations based on probability and chance the experts attempt to show how minimal a specific risk is.
3 Showing wise forethought for future needs or events.
4 This situation relates to a fundamental principle of insurance called the Law of Large Numbers which states that, 'the larger the group of similar exposure units, the more closely the actual losses experienced by that group will approach those that can be anticipated...[it] means that the greater the number of exposure units, the more accurate the insurers can be in calculating their premiums. This is because they are better able to assess the size of future loss payments and hence to work out an appropriate charge that will enable them to cover those losses' (Diacon and Carter, 1995, p.3).
5 Redlining has also been studied in relation to the lending practices of banks, mortgage firms and loan institutions (see for example Dingemans, 1979; Kantor and Nystuen, 1982; Jones and Maclennan, 1987; Engels, 1994; Perle, 1994; Guskind, 1995; Dymmski, 1995; Cho, 1996; Li, 1996; Holmes *et al*, 1996; Tootell, 1996). The behaviour of these institutions is, however, often influenced by the insurance industry, as lenders will not be keen to do business in areas where insurance cannot be obtained and which are seen as 'risky' or what Beck referred to as 'loser regions'.
6 Since hazards and postcodes do not share the same boundaries it is difficult to construct a policy that precisely reflects the risk exposure (see for example Raper *et al*, 1992). This pattern is further complicated as insurers often disagree as to which areas are high or low risk or in relation to the magnitude of the risk.
7 After the riots, the Secretary of State, Michael Heseltine, called upon the private financial sector, including insurers, to help improve conditions by supporting government efforts to regenerate the inner-city (Falush, 1994). It was alleged that redlining was blocking businesses in the inner city. The Association of British Insurers (ABI) in conjunction with the Government set up a task force to look at these allegations. The ABI would not however guarantee to provide policies at realistically affordable rates or without heavy security protection (Threadgold, 1995). This

effectively stopped established businesses from re-opening and new businesses from locating in these stigmatised areas. Insurers effectively blacklisted inner-city estates (Murray, 1994).

8 For example, the City of London Police defined terrorism as 'the deliberate use of violence and threat of violence to evoke a state of fear (terror) in a particular victim or audience. Usually the use and threat of violence are directed at one group of targets (victims) while the demands for compliance are directed towards a separate group of targets'. This definition was developed during the early 1990s when the City Police were responsible for drawing up a counter-terrorism strategy for the Square Mile. This definition encompasses the three 'cornerstones', which can be seen to shape the geography of security in the City of London. Specifically it paints a picture of unsuspecting victims (businesses in the Square Mile) who are exploited by the objectives of terrorism. However, this definition is subjective and value-laden coming from a policing perspective.

9 The Provisional IRA's political wing.

10 For example he points to the way in which terrorist organisations in Northern Ireland attempted to undermine confidence in the police and army by constantly referring to them in the media as 'thugs in uniform' (p.39).

11 For example, on 24 June 1939 the IRA were responsible for a series of attacks against six separate banks in London.

12 The controlling council of the Provisional IRA.

13 Daithi O'Connaill, interview with *Weekend World* (London Weekend Television), 17 November 1974.

PART II
THE CITY OF LONDON'S RESPONSE

Chapter 5

Constructing and Reinforcing the Ring of Steel

Introduction

The City of London has always been a defined and contained territory which has required defending at various stages in history, either against crime or intrusion or, against economic competitors. The response to the terrorist threat since the early 1990s combined with the need to remain competitive in the global economy was the most recent example of this trend. Following a brief introduction to the built form of the City, this chapter will chart the physical changes to the urban landscape, and the associated strategies employed by the agents of security, primarily the City of London police, that were developed in the Square Mile as a result of the threat of terrorism in the 1990s and early twenty-first century.

Recent attempts to defend the City from physical attack is by no means the first time the landscape of the Square Mile had been fortified to protect itself and maintain its economic pre-eminence. The City of London was formed in 43 AD as a Roman military base centred on a rectangular fort and steadily developed into a leading commercial centre in the preceding centuries.[1] The site was in a good defensive position with rivers on the South (Thames) and West (Fleet) and lower lying ground to the North and East. The City was first 'walled' in about 200 AD when a substantial defensive stone wall was built around the City on the landward side.[2] Enclosing the City from the site of the Tower of London in the East to Blackfriars in the West, this main wall incorporated the north and west wall of the earlier fort, with an outer ditch and a supporting bank of earth against the inner face of the wall completing the defences. The defences were further enhanced towards the end of the 4th century AD by the construction of a riverside wall and by the addition of bastions to the outside of the landward wall in the east of the City. Entrance to the City was gained through a number of gates in the wall.[3] The City of London at this time could be seen as a classic example of a stronghold as described in Chapter 2. In time, the function of the City of London changed, from initially that of a fortress, to that of a port and commercial centre.

Following a period of Roman withdrawal, Alfred the Great in the ninth century repaired the walled defences creating a fortified garrison town. This wall subsequently formed the boundary of the City of London, which continued to develop into a domestic and international commercial centre. The remains of this wall are still visible today. At this time markets were created for insurance, shipping, and commodities, which laid the foundations for the City's future economic success.

The Buildings of Global Finance

The trading links between the City and the rest of the world continued to expand especially in the sixteenth century with the development of new trade routes to the east, and the discovery of the Americas. London developed during this time into a mercantile city through the expansion of ocean trade, and by 1700 was the largest city in the world with the City playing the pivotal role in this expansion (Duffy and Henney, 1989; Corporation of London, 1995). This expansion continued, allowing the City to become the world's leading financial centre between the Franco-Prussian war (1870-1) and the First World War (see for example Cassis, 1985a, 1985b; Harris and Thane, 1984; Lisle-Williams, 1984). This coincided with a period of history in which Britain dominated international trade with the City becoming the 'central switchboard' for the increasingly global market (Anderson, 1987, p.24). As the industrial revolution progressed in Britain, the growth of manufacturing highlighted the demand for the City's commercial and financial expertise, which was adapted to fulfil its role as a nerve centre of the Victorian empire. This was reflected in the built form of the City with 'its grand buildings' standing as 'a testament to the City's historic centrality' (Jacobs, 1994, p.751; see also Daniels, 1993; Jacobs, 1993).

The inter-war years saw the partial erosion of the City's pre-eminent position in the world economy through increased competition from America. This competition continued and intensified in the 1950s and 1960s with further competition from America as well from Western Europe and Japan. However, as Jacobs (1994, p.751) noted, 'in the span of half a century the City of London had gone from the centre of an empire with global reach, to one of three urban centres given the privileged designation of global city' (see also Pryke, 1991; Sassen, 1991; Thrift, 1994). This was reflected in its built environment with the increasing development of buildings of global finance.

In the 1980s the City was 'both a postimperial city and a postmodern(ising) city: it was a city of transition and change' (Jacobs, 1994, p.751). The contemporary role of the City in the global economy necessitated periods of intense building development, the most recent of which, in the late 1980s, led to the construction of 'postmodern' office complexes, which became the new status symbols of the City, advertising the wealth and power of the Square Mile. Their development can be linked to the deregulation of financial markets in London in 1986 (the so-called Big Bang), which necessitated new types of buildings appropriate for multinational financial companies.[4] Indeed, in the mid-1980s and early 1990s one-third of the City's buildings were rebuilt or replaced as competition from other financial centres intensified. As Cathcart (1993, p.19) reported:

> In the mid 1980s, after a long period of restriction on new development, the Corporation saw that the electronic markets that followed Big Bang would require new trading floors and new kinds of building. It looked east and saw just such buildings being planned in Docklands and other European financial centres. If the City did not open the way to change, it concluded, it would be abandoned...The result was spectacular: in eight years, one-third of all the buildings in the City were replaced.

In short, during the 1980s and 1990s the City of London underwent a significant rebuilding cycle linked to reinforcing its position in the global economy primarily by the construction of a host of new buildings which could cater for the demands of modern business. Modernist architecture became antique as the City increasingly renewed itself (Daniels and Bobe, 1992). Despite the slowing down of the building boom in the 1990s, individual projects are still being put forward in an attempt to express the character of the City.

Whilst noting the necessity of these new buildings, the City was also concerned with maintaining its heritage identity. This commitment was reinforced by the Local Plan of 1986 which highlighted that preservation of the City's historic character was important to its continual role as a financial leader. As the Corporation of London (1986, p.3) noted:

> ...the City... is noted for its business expertise, its wealth of history and special architectural heritage... [giving it]... a world-wide reputation... and distinguish[ing] it from other international business centres. The... City's business activities, which are underpinned by the benefits of its precious heritage, further the wealth and opportunities of London and the surrounding region, and also provide a significant contribution to the well being of the nation's economy.

This all served to make the Square Mile in the 1990s a unique place of significant tradition and symbolic importance, and one which helped promote the City as 'the place' to conduct business. Paradoxically, it also made it a key terrorist target, given the concentration of high value properties owned by global institutions, which would guarantee significant media attention if attacked. The remainder of this chapter will analyse how the reaction to this terrorist threat led to the development of anti-terrorist security in and around the City which sought to control and regulate space, reinforcing localisation whilst still attempting to enhance the global economic function of the Square Mile.

The Evolution of the City's Defensive Landscape

In the early 1990s the City was attacked a number of times by the Provisional IRA, although those responsible for security in the Square Mile also saw the threat of bombing as coming from terrorist groups other than the Provisional IRA (see Table 5.1 below).

Table 5.1 Provisional IRA incidents in the City of London in the early 1990s

20/7/1990	A bomb explodes in the Stock Exchange
29/2/1992	A device explodes at the Crown Prosecution Service in Furnival Street
10/4/1992	A large van bomb explodes outside the Baltic Exchange in St. Mary Axe
25/6/1992	A device explodes under a car in Coleman Street
24/4/1993	A large vehicle bomb explodes in Bishopsgate
28/8/1993	A device is recovered from Wormwood Street near Bishopsgate

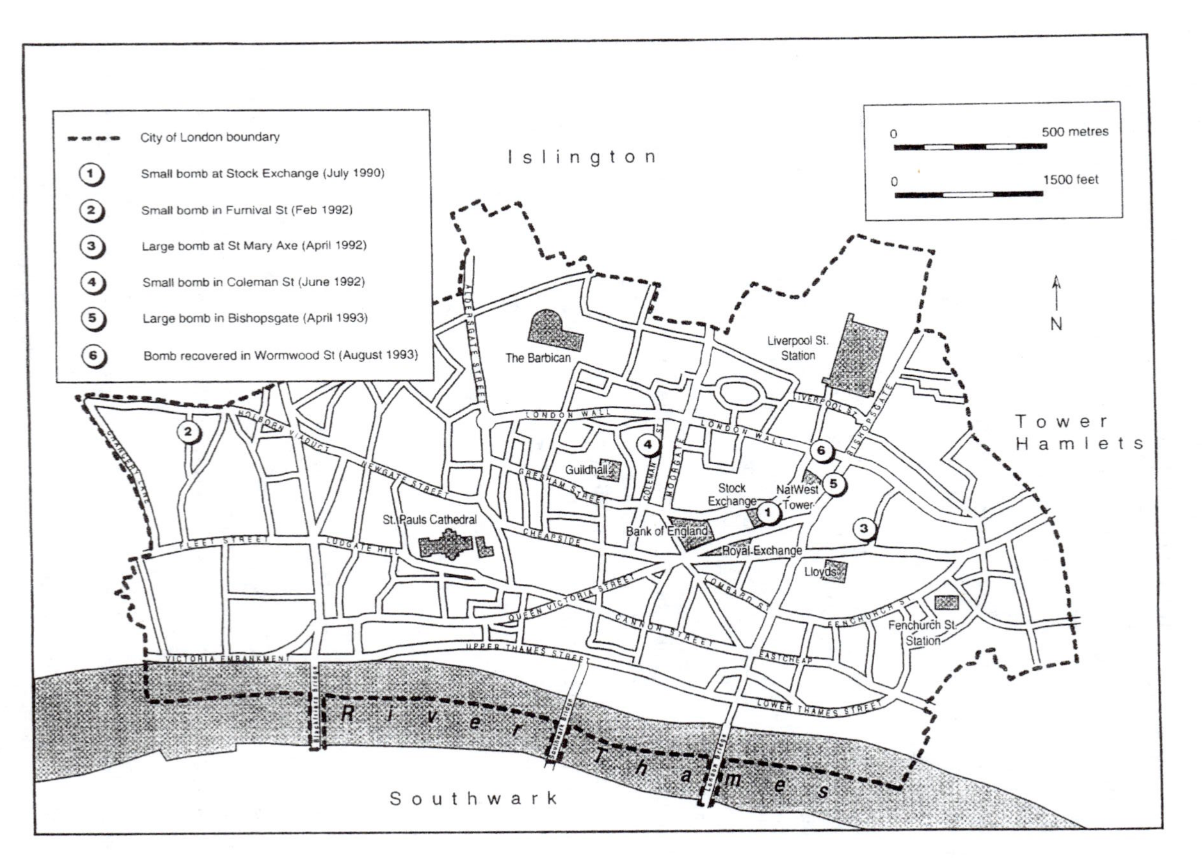

Figure 5.1 Map of Provisional IRA incidents in the City of London in the early 1990s

The City was attacked by the Provisional IRA not to cause major loss of life (the two main bombs in April 1992 and April 1993 were detonated at a time when the City was virtually deserted), nor to impact upon the concentrated transport infrastructure to be found in the Square Mile. Instead, it was attacked to cause economic disruption (in particular through insurance claims) and to put additional political pressure on the British Government to remove themselves and their troops from Northern Ireland.

As a result of the terrorist threat a variety of attempts were made to 'design out terrorism' through the construction of a number of defensive features and modifications made to urban design. These attempts can be characterised in eight stages between 1990-2003, which brought about distinct changes to the physical landscape (see Table 5.2). The agents of security adopted a series of territorial strategies that gradually became more advanced, and directly impacted upon a greater geographical area both within and outside the boundaries of the Square Mile. These stages therefore relate not just to the direct physical changes that took place but also to the management strategies employed by the agents of security, most notably the City of London Police, who were responsible for activating the major security measures. In short, over the past decade the Square Mile, the City has increasingly been separating itself from the rest of London in both physical and technological terms (Coaffee, 2000a, 2000b, 2002; Power, 2001; Graham and Marvin, 2001). It will also be highlighted that in the City the extent to which the security apparatus was fully mobilised was related to the perceived threat level which ebbed and flowed through since the early 1990s, as well as the impact of non-terrorist events such as anti-capitalism demonstrations and the development of traffic management schemes for central London.

Apprehension

As highlighted in Chapter 4, the Provisional IRA's main bombing campaign in the 1990s was aimed at economic targets in London with one of the first attacks occurring at the Stock Exchange in the centre the City of London. This bomb exploded in the public gallery causing much damage to the visitor area. No one was injured due to a telephone warning being received. Of perhaps more political significance however was the mortar bomb attack on Downing Street on 7th February 1991, which symbolically carried the Provisional IRA's message to the heart of the establishment in an attempt to force the British Government into political dialogue.[5]

After the Downing Street attack the Provisional IRA began to extend their campaign to cause maximum civilian disruption, which included a number of attacks against transport facilities. Such attacks included a litter bin bomb at Victoria Station, which killed one person and led to the bins on the London Underground being removed to prevent similar incidents. Perhaps aware of the state of the developing recession in Britain, and the pressure on government finances, the Provisional IRA at this time also began to appreciate the value of inflicting massive economic damage on Britain, although this may have been more a question of stumbling upon

Table 5.2 Stages in the evolution of the ring of steel

Stage	Dates	Key Incidents	Main Features of Response
Apprehension	1990 – April 1992	• Beginning of IRA mainland attacks against economic targets (Stock Exchange July 1990) • Furnival Street bomb (February 1992)	• More overt policing especially for major events • Plans for City-wide security schemes
Containment	April 1992 – April 1993	• St. Mary Axe bomb (April 1992) • Colman Street bomb (June 1992)	• Armed Police checkpoints • Traffic management enhanced • CCTV adapted for counter-terrorism uses
Deterrence	April 1993 – Sept 1994	• Bishopsgate bomb (April 1993) • Business lobbying for enhanced security • Wormwood Street bomb find (August 1993)	• Security checkpoints introduced • Advanced police-operated CCTV and alert systems • Private CCTV schemes • Increased no-parking areas
Optimism	Sept 1994 – Feb 1996	• Provisional IRA ceasefire	• Downgrading of visible police presence • Checkpoints become permanent • Updated police-operated CCTV
Reactivation	Feb 1996 – Feb 1997	• Docklands bomb (February 1996) • Subsequent attacks in London and Manchester	• Large increase in visible policing • Increased frequency of roving checkpoints • Use of legislation to increase stop and search
Extension	Feb 1997 – June 1999	• Decision to extend the ring of steel westwards	• Expansion of ring of steel coverage • Advanced CCTV employed • Environmental improvements highlighted
Re-appropriation	June 1999 – Sept 2001	• May Day/ anti-capitalism riots (June 1999) • Subsequent anti-capitalist demonstrations • Threats from dissident Irish republican terrorists	• Proactive enhancement of private building security • Better liaison between City police and other Forces • New anti-terrorism legislation
Reappraisal	Sept 2001 – present	• 9/11 and fear of further attacks • Anniversary of 9/11	• Re-examination of counter-terrorism procedures • Increased uptake of alert systems • High state of alert especially on 'key dates'

a strategy. For example, they would have noticed that in late 1991 and early 1992 the number of large bombs they detonated in Belfast city centre caused massive costs, both in terms of actual damage caused (with no injuries), and insurance and compensation claims, as well as negatively affecting Belfast's identity as a place of business. This concept of economic disruption was then put to use in London, and in particular against the City.

However, the perceived threat level before the 1992 bomb in the City was not considered high enough for the police to establish any special anti-terrorist measures. Before the St. Mary Axe bomb the City Police would offer general security advice when approached but had not at this point set up any specific anti-terrorist campaigns or altered the physical landscape in any way.

> Prior to the first bomb we would offer advice but generally there was not a concerted campaign as such as the City had only been targeted in 1990 at the Stock Exchange and that was only a small device
>
> (Senior Police officer)

As 1991 progressed the perceived threat to the City from the Provisional IRA was enhanced as bombs continued to explode in London. As such the City Police increasingly felt the need to defend the Square Mile from terrorism. In his Annual Report (1991) the Commissioner of the City Police referred to the question of likely terrorist attacks by the Provisional IRA, given other attacks elsewhere in London. He noted that 'although we had no serious incidents in the City in 1991, the effects of terrorist attacks elsewhere in London had a significant effect in heightening the need for even greater security… (p.5)'. In particular the potential for attacks on the City's transport infrastructure was noted given the five mainline railway terminals, City Thameslink, and the ten underground stations in the Square Mile. In response the police significantly enhanced security arrangements for major events in the City, for example through the deployment of a specialist counter terrorism search team. Such high profile policing, the Commissioner noted, had also led to a ten per cent reduction in recorded crime.[6] Thus for the first time a counter-terrorism effort was having a noticeable knock-on effect on other areas of criminal activity. It will be shown later in this chapter how this effect, in time, became one of the central justification for maintaining an overt anti-terrorist strategy.

During the early months of 1992 a number of bombs were planted in and around London to coincide with the forthcoming General Election, which aimed to keep the Northern Ireland issue at the top of the political agenda. Election arrangements thus determined the timing of the Provisional IRA's strikes, whilst economics determined the target of their attacks. In early 1992 over 15 bomb hoaxes were received directly relating to the area covered by the City[7] and in February a small terrorist bomb exploded in Furnival Street in the north-east of the City. No substantial damage was caused but two police officers were treated for shock. The police were now beginning to take the threat to the City more seriously, although little in the way of proactive security enhancement was contemplated until the threat level was further enhanced in April 1992 by the bombing of the Baltic Exchange at St. Mary Axe in the heart of the Square Mile.[8]

The period prior to April 1992 therefore saw a heightening of the threat level faced from terrorism and the beginnings of a territorial approach to policing strategies at a scale not previously seen in England. With the Provisional IRA increasingly targeting the mainland, and in particular at this time commercial and transport infrastructure, the City began to develop strategies to reduce the threat. This was a particularly *ad hoc* approach and was generally only undertaken during high-profile events occurring in the City. The ever-changing strategy of the Provisional IRA at this time also made it difficult for the police to justify, on cost grounds, significant and permanent landscape alteration to limit access and enhance surveillance opportunities, when the City was considered only one of many potential targets in England. In particular, overt security in the City would have been politically unacceptable at this time.

Containment

The St. Mary Axe bomb exploded on the evening of 10th April 1992 on the day of the General Election. Shortly before 9pm on the evening of 10 April 1992 a large van-bomb exploded outside the Baltic Exchange on St. Mary Axe. During this attack 3 people were killed and over one hundred injured as well as a large area being devastated. A coded warning was given but, according to the Police, the caller identified the Stock Exchange as the target. This was consistent with the targeting that had gone on over the election campaign. Immediately after this bomb emergency plans were devised to prevent further attacks. This was the first major bomb in the City, and it was felt that an increased police presence, with officers carrying out spot checks on vehicles, was going to be a suitable response:

> After the 1992 bomb we [the Police] were still really only geared to offering advice, but we did create a much higher profile of policing on the streets... but the sort of things we did then were not as structured and co-ordinated as they were after 1993.
>
> (Senior Police officer)

It was not considered appropriate at this time to use advanced security techniques to defend the City. As the leader of the Corporation of London's Policy and Resources committee, Michael Cassidy, noted:

> A ring of steel solution didn't arise then because it was the first such incident. I think people were taking the view that the policing was going to be very front-line and vigilant.

Therefore, putting more officers on patrol was seen as the best operational response. To increase their manpower on the street, the City Police took many officers away from desk jobs and allocated an additional £1.5 million to recruit an extra forty police officers in order to increase patrols within the Square Mile. In total over ninety more police officers were able to patrol the City streets, and campaigns to recruit and train a number of special constables were initiated. Strategically, the police, stretching the Police and Criminal Evidence Act (PACE) to its limit, instigated a number of short term 'roving police checkpoints'[9] on the major entrances into the City with armed officers. These began operating on 30 November 1992 and have been operational ever since. In addition to the random road checks a special team of officers known as the Counter Terrorist Search Team (CTS) was trained and subsequently deployed on nineteen occasions during 1992. Searches were particularly thorough before important ceremonial occasions such as the Lord Mayor's show.[10]

However, in May 1992 a month after the St. Mary Axe bomb, an evaluation report presented to a number of Corporation of London committees agreed that Local Plan and draft Unitary Development Plan (UDP) for improving the City's environmental and movement policies called 'Key to the Future' should be implemented. Under these proposals drawn up *before* the 1992 bomb it was planned to pedestrianise a central part of the City around the Bank of England and alter traffic signalling on others, to improve traffic flow and reduce pollution.[11] These policies, as will be shown, became fundamentally important in the City being able to construct anti-terrorist security measures in the preceding months and years. For example, an evaluation report of this was presented to the Traffic Management and Road Safety Sub-committee on 13 April 1992 just after the bomb, highlighting that the Corporation's movements and transport policies in the UDP aimed to: reduce the impact of through traffic; remove excess traffic from local roads; improve safety for highway users and pedestrians; and minimise noise and atmospheric pollution.

As a result, after the St. Mary Axe bomb, with the support of the Commissioner of Police, some traffic management measures in line these proposals were introduced on an experimental basis on three City roads which could be carried out under Section 9 of the Road Traffic Regulation Act, 1984:

> The... experimental schemes... will each contribute to the implementation of the Corporation's policies in the local plans in the interests of pedestrian safety and the improvement of the City's environment.
>
> (Minutes of the Court of Common Council,[12] 4 June 1992)

No official reference at this time was made to the anti-terrorist security implications of these changes as the Corporation were keen to downplay the impact of the St. Mary Axe bomb. It was also noted at the Policy and Resources meeting on 18 June 1992 that the proposals for Red Routes (no stopping by vehicles allowed at any time) were being considered for implementation in the City. These were also capable of helping the security effort by reducing the areas in the City where vehicles could park. Furthermore, waiting restrictions on a number of roads in the vicinity of the Old Bailey, the Bank of England, the Stock Exchange and the Lloyds building were extended. This was a direct recommendation of the Commissioner of Police, who used his powers to suspend metered parking around these areas. This affected a total of fifteen roads around these buildings.[13] These minor modifications, together, resulted in a heightened awareness of the dangers associated with unattended vehicles (without distinguishing features) on City streets.

The restrictions on movement around the central core of the City can be viewed as the first attempts by the police to single out a particular spatial area of the City for special security attention. These measures were officially done to ease emergency vehicle access but can be seen as an attempt to defend the highest risk targets in the City (i.e. the most prestigious landmark buildings) from potential terrorist attack.

The police at this time wanted to set up permanent vehicle checkpoints on all entrances to the City, which they felt could be undertaken in line with 'Key to the Future' policies. However, a combination of legal (related to PACE), financial restrictions (primarily that that City businesses would not want to pay for such measures) and, public opinion, made this impossible to do at that time. First there was a feeling that a permanent security cordon would be an over reaction given that this was the first major bomb in the City. Second, there were fears that such permanent arrangements could have made the City look like Belfast, giving a propaganda coup to the terrorists (Kelly, 1994a). Consequently, a heavily fortified environment, overtly highlighting the extent of the terrorist threat, was seen at this time as potentially counterproductive in reducing the fear of terrorism. However, minor changes to the landscape did occur at this time including the removal of over one thousand litter bins that could not be made bomb-proof and, reducing the time black refuse sacks were left on the street. These changes were enacted to reduce the number of places a bomb could be concealed.

However, a minor explosion in June 1992 in Coleman Street further enhanced the calls for improved anti-terrorism measures to be constructed. This bomb, although only a small device, showed that security in the Square Mile could be easily breached. As a result there was a rising in awareness of the threats faced by City occupiers.

In addition to random road-checks and access restrictions, the coverage of traffic management CCTV (Area Traffic Control - ATC) was extended and adapted to focus on incoming traffic whilst private businesses were encouraged by the police to install CCTV cameras as another deterrent:

> After the 1992 bomb there was an appreciation that CCTV could play a part in combating crime generally, as well as terrorism specifically, as the terrorists had been caught on film.
> (Senior Police officer)

He continued by indicating that the feeling at this time was that the police must not just sit back and wait for the Provisional IRA to act, but must be increasingly proactive in trying to combat the threat. In particular, the police were beginning to liaise with private businesses and their security personnel as to how they could best cope with the ongoing threat.

The period of containment after the St. Mary Axe bomb can be assessed as a time when there was an increased perception of the terrorist threat, and hence strategies were developed to enhance the control of space, limiting the ease with which terrorists could move around the City, and ensuring that vehicles parked suspiciously could be easily identified. The increased use of access restriction, police patrols and the modification of CCTV systems were, however, highly focused within a central core area of the City containing the most high profile institutions. The extent to which territorial control strategies could be introduced at this time by the police was limited on legal and financial grounds and by the fear that overt defensive landscape features could lead to the City becoming stigmatised and a less attractive business location. It was felt that this could lead to a situation where 'fear would follow form' as it did in Belfast in the 1970s with heightened security leading to an increased sense of danger being associated with the central area.

Despite the proactive management strategies adopted by the police there was still a tremendous sense of apathy from sections of the business community with regard to the security threat, indicating that the risk was not accepted as realistic:

> St. Mary Axe was a watershed as far as anti-terrorist measures were concerned - it focused the minds of people on what could be done. This advice was taken up by some but others held the view that lightning doesn't strike twice and failed to heed the warnings.
>
> (City security advisor)

He continued by indicating that this view began to change after the 1993 bomb:

> Bishopsgate in 1993 was therefore much more significant than St. Mary Axe as it proved this wrong.

Deterrence

In the months preceding the Bishopsgate attack much correspondence occurred between the City of London Police, the Royal Ulster Constabulary (who looked after Belfast's security cordon), security experts, and transport engineers in order to develop a blueprint for a possible ring of steel for the City. This was to be based, in part, on historical experience in Belfast and other towns in Northern Ireland that had previously developed similar security cordons. The nature of this collaborative relationship between the two police forces will be further discussed in later chapters. It is also important to note that the week before the Bishopsgate bomb a survey was carried out by the Corporation of London to reveal the likely impacts of the major traffic and environmental changes that were planned in line with the 'Key to the Future' policies that were due to be carried out in the preceding six weeks (Power, 1993). These were seen as supportive of, and in many cases synonymous with, security enhancement.

Meetings with the RUC and consultation with regard to traffic and environment policies suggest that after the 1992 bomb, the City Police were in the process of assessing the potential applicability of introducing security measures to control access into selected parts of the Square Mile well before the 1993 bomb. Given the concerns of City occupiers who wanted to avoid being associated with the 'stigma' of an anti-terrorist cordon, the police and the Corporation of London were attempting to bring in centralised strategies of security under the guise of environmental and movement policies, which were seen, at this time, to be politically acceptable. These will be detailed in Chapter 7.

Evading the Post-1992 Security Measures

In hindsight, despite the road checks and other measures set up after the St. Mary Axe bomb, the police were well aware of the limitations of PACE which prohibited permanent checks on vehicles coming into the City. By using mobile communications, and a scout ahead to tell them whether or not there was a police check operating, terrorists could easily plant a bomb in the City.[14] During the early months of 1993, with the delicate state of political dialogue aimed at brokering a Provisional IRA ceasefire, the risk of further attack to the City had significantly increased. There were a number of factors which contributed to the increased threat level from the Provisional IRA. First, there was a lack of progress with political talks. Second, it was felt the Provisional IRA had more than sufficient funds, equipment and personnel to launch a sustained attack in England. As such a warning circulated to all police forces on the mainland highlighting intelligence reports that suggested the Provisional IRA might strike:

> I heard on the grapevine that a major strike against the City was a possibility due to the failing of political talks at that time. I know the City intensified their road-checks at this time but they failed to intercept the lorry-bomb as the IRA simply exploited the inherent weaknesses of the random checks on traffic by the Police.
>
> (City-based Risk Manager)

After a period of political negotiations between Sinn Fein and the British Government had failed to achieve any progress, the worst fears of the police were realised on 24 April 1993 when a Provisional IRA bomb exploded in Bishopsgate in the east of the City.[15] The location of this bomb was away from the more highly protected core in the centre of the City (see Figure 5.1), which had been subject to additional access restrictions to counter terrorist attack after the 1992 bomb. The edition of *An Problacht/Republican News*[16] on 29 April 1993 gave the Provisional IRA's version of events and highlighted how 'having spotted a breach in the usually tight security around the City' they planted their bomb and issued several warnings. They further noted how their 'surveillance operatives' exploited a loophole in security, which allowed builders' vehicles to park on the double yellow lines on Saturday mornings without being searched or asked to move by the Police:

> IRA surveillance on the targeted site informed them [the bombers] that on Saturday morning's builders' vehicles are allowed to park on double yellow lines to carry out repair work. Most of this repair work, ironically enough, was being carried out after last year's massive Baltic Exchange explosion... Despite heavy security the IRA volunteers on board the truck bomb took the decision to move into position. Bypassing a number of police units, they parked their vehicle into its pre-selected spot immediately adjacent to the Hong Kong and Shanghai Bank.

It was well known that close to this site, at this time, the European Bank of Reconstruction and Development had been meeting. Through the precise placement of this bomb the Provisional IRA succeeded in making sure the impact of the bomb was felt politically as well as economically. As the Provisional IRA further noted:

'as well as the huge cost of structural damage, the loss of buildings and the knock on effects of insurance costs, the City of London is assessing the damage to its prestige as a world financial centre.'[17]

The Reaction to Bishopsgate

Immediately after this bomb the Corporation's Chief Planning Officer called for the demolition of the NatWest Tower and other buildings damaged by the blast. He saw the devastation as an ideal opportunity to rid the City of London of some of its 1970s architecture constructing a new state-of-the-art structure as a 'symbol of defiance to the IRA'. The Corporation of London, however, strongly distanced themselves from his comments, noting that the NatWest Tower was an integral part of the City's skyline.

In the wake of the bomb the media, and sections of the business community, also began to suggest that drastic changes should be made to City security (Jones, 1993). An editorial in the *Sunday People* the day after the bomb captured the popular view that security must be enhanced in the City:

> If we are to wage effective war against the IRA, there must now be an urgent review of security at their most likely target. Since the IRA mortar-bombed Number 10 from a waiting van, nothing is allowed to park in Whitehall. IF IT CAN BE DONE FOR DOWNING STREET IT CAN BE DONE IN THE CITY.
>
> (25 April, p.2, emphasis added)

Leading City figures cited in *The Times*, also indicated that a Belfast-style scheme should be implemented:

> The City should be turned into a medieval-style walled enclave to prevent terrorist attacks... (p.3) In private there is talk about a "walled city" approach to security with access through a number of small "gates" and controlled by security discs...
>
> (27 April, p.27)[18]

In a similar vein, *The Sun* newspaper published a six-point plan 'which would keep the IRA bombers out' but only at great cost. They proposed a 'ring of steel' based on the Belfast model, which they thought would cost £100 million to implement plus £25 million a year to run. Under this proposal they suggested that entry be restricted to 8-10 entrances, which were fitted with security barriers manned by armed guards. These barriers they proposed would only operate during the evening between 7pm and 7am and at weekends, as they would be impractical to operate during working hours. During office hours only valid business traffic would be permitted entry. Other security methods this article suggested were an additional 250 police officers on the streets of the City, which was estimated to cost £5 million a year, and a lorry ban in the City. The article also suggests that a national identity card scheme, a new National Anti-Terrorist Squad and bomb proofing of vulnerable buildings could also help prevent attacks (see Kay and Lewthwaite, 1993).

Other views differed with some highlighting the pitfalls of high levels of overt security – the 'fear follows form' argument. For example the City engineer, evoking medieval metaphors, stated that 'we wouldn't want the City turned into a castle with a moat around the outside'.[19] At this time such 'draconian security was dismissed as a propaganda gift to the Provisional IRA as well as being difficult to implement legally'.[20] Initially the leaders of the Corporation wanted to take stock of recent events before deciding whether or not to implement such radical proposals. As Michael Cassidy, referring to the day after the bomb, noted:

> It's true that on that Sunday morning my reaction was that this is such a radical proposal that I couldn't immediately say it was one that we could embrace and support.

This led to a series of talks between the Corporation, the Lord Mayor of London and the Government on ways in which the City should defend itself. In May 1993, given the heightened risk of further attack, and severe pressure from the business community, (especially the foreign institutions) to improve security (see Chapter 7) the police confirmed that they were considering radical plans, in the form of a security cordon, to deter terrorists from the City. The construction of the proposed scheme was essentially the same strategy that was being suggested on movement and environmental grounds prior to the Bishopsgate bomb. The emphasis of the rationale behind the scheme had however taken on a very different public persona, being seen as a direct response to the Bishopsgate blast rather than the move towards environmental improvements (the latter would re-emerge, in time, as the City's main justification for the ring of steel). Opinions on enhanced security had moved from a position where they were rejected after the 1992 bomb – fear follows form – to a reversal of this position where changes in the form of the landscape were seen to follow the fear of further attack – form follows fear.

The police had been consistently calling for greater security measures, and had reached the conclusion that permanent road-checks, ensuring that all vehicles entering the City were captured on camera and were potentially liable to be searched, was the only way they could stop another bomb. Legally this was not possible after the St. Mary Axe bomb. Under PACE (1984), police are only allowed to search vehicles under suspicion of terrorism. The police believed the Act left them open to civil action in cases where a road check does not lead to a conviction (Burns *et al*, 1993).[21]

However after Bishopsgate it was seen as better to risk abusing police powers under PACE by setting up a fixed cordon, rather than experience further bombings. As the Commissioner noted in the days after Bishopsgate: 'I want the power to set up road checks wherever and whenever, without specific reason for doing so'.[22]

Initially it was thought at the time that the security cordon could be in place within a year. However, the seriousness of the situation, given the risk of further terrorist attack, meant that construction of the so-called *ring of steel*[23] began in June 1993, with the removal of all remaining litter-bins, and the introduction of security checkpoints to bar all non-essential traffic.

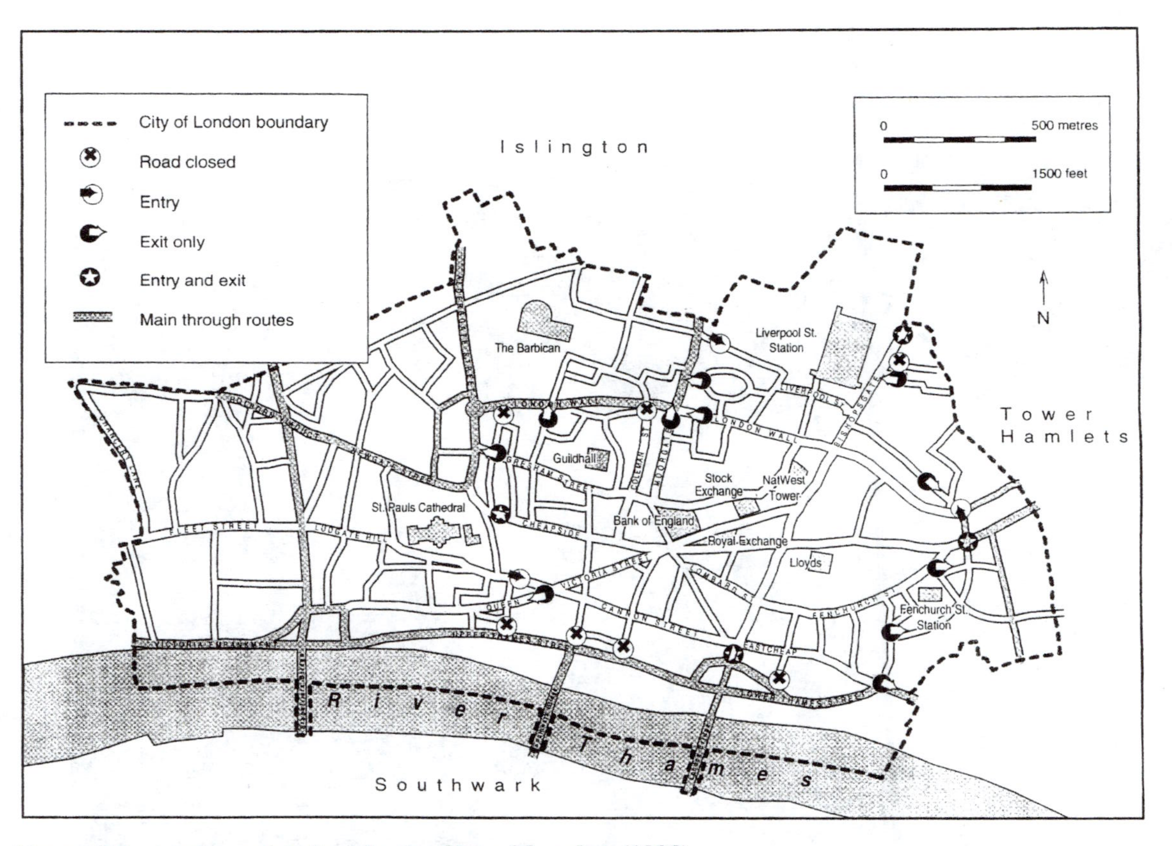

Figure 5.2 Access restrictions in the City of London (1993)

Figure 5.3 Entrance into the 'ring of plastic' (I)

Figure 5.4 Entrance into the 'ring of plastic' (II)

Such modifications were criticised by those who felt that it would cause traffic chaos at the boundaries of the City. Some felt that checkpoints could lead to large queues when entering the City as well as vehicles being increasingly pushed into neighbouring boroughs. There were also fears that such a radical scheme could geographically displace the risk of terrorist attack to other areas as it had done in Belfast. Kenneth Clarke, the Home Secretary at that time, noting the economic importance of the City, summed up the situation the City faced, indicating the delicate balancing act between security and business normality that was required:

> There is a balance to be struck between having roadblocks which will frustrate what the terrorists can do, and creating enormous traffic jams which would disrupt the life out of the City.[24]

Eventually on the weekend of 3-4 July 1993 a full 'Belfast-style' ring of steel was activated in the City, securing all entrances. The main access restrictions imposed are shown in Figure 5.2. This shows that most routes were closed or made exit only, leaving seven routes (plus one bus route) in which the City could be entered. On these routes into the City, road-checks manned by armed police were set up (Figures 5.3 and 5.4). Locally the ring of steel was often referred to as the 'ring of plastic' as the temporary access restrictions were based primarily on the funnelling of traffic through rows of plastic traffic cones, as the scheme was still officially 'temporary'. The City's ring of steel looked nothing like that constructed in 1970s Belfast. It represented a far more symbolic and technologically advanced approach to security, which avoided the 'barrier mentality' in favour of less overt security measures. However, the 'ring of plastic' provided a highly visible demonstration that the City was taking the terrorist threat seriously, even if many entering the City did not realise its anti-terrorist use.

The ring of steel did not provide full geographic security coverage within the Square Mile, although the area it covered was in excess of the area in which traffic modifications were established in 1992.

As such, much of the western side of the City still remained outside the official cordon (see Figure 5.5). The cordon was positioned in this way for a number of reasons related to the spatial configuration of the existing urban landscape. Initially, the police Commissioner wanted to make sure that all the key financial targets were included, leaving some businesses unhappy about the exact placement of the cordon as it left them outside the secure zone:

> Some people were discontented that they were excluded. The point is that we had to begin and start somewhere, and, like it or not, there is a part of the City that is more vulnerable and has a greater economic value to the nation and the international economy as opposed to just the City of London, which, is why the inner cordon was focused on that particular area at the outset.
>
> (Senior Police officer)

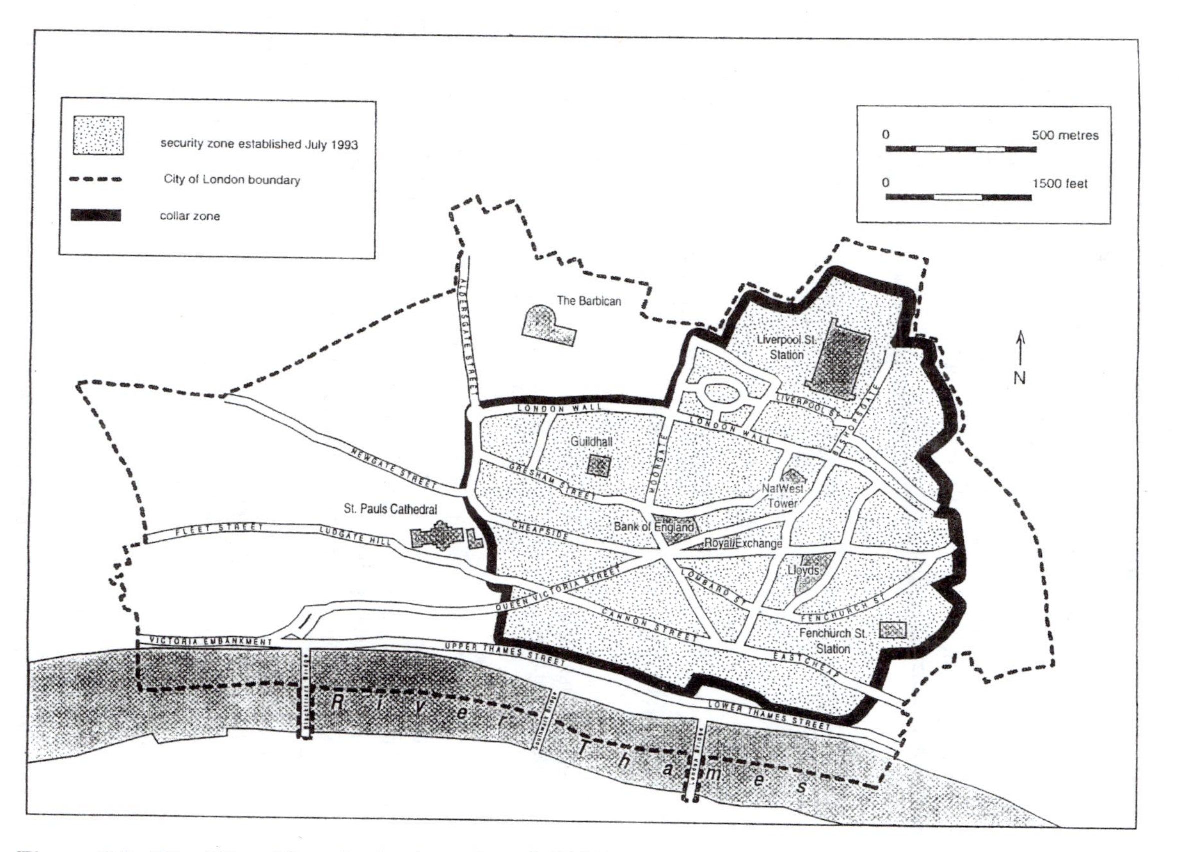

Figure 5.5 The City of London's ring of steel (1993)

The ring of steel was an enhancement of prior strategies to secure this area, but was also developed so the traffic flow through the City stayed the same. The ring of steel, or 'Experimental Traffic Scheme' as the Corporation officially called it, was put in place for an initial period of six months with the possibility of a further six months extension. After a maximum of twelve months the Corporation of London had to apply for permission to make the scheme permanent. This involved consultation with neighbouring boroughs and other agencies that might have objections. This meant that throwing a cordon around the entire City would have led to all traffic being diverted to neighbouring boroughs leading to impacts outside the City's spatial jurisdiction. Politically these would have been untenable given the array of predominantly Labour controlled borough councils which encircled the City:

> Logistically it was a nightmare to identify which streets you were going to put blockers on and which way you were going to have traffic going because we had to reverse the one-way streets in some areas.
>
> (Senior Police officer)

To help them the Corporation utilised computer programmes that could predict traffic-flow in the City.[25] As Michael Cassidy noted:

> This programme could predict what would happen if we closed certain streets. For the process of defining the cordon we started at the very outer borders of the City and found that traffic disruption would have been unacceptable. What we did then was to shrink the cordon to a point where the knock on effects we felt were just about acceptable.

The police in effect wanted to 'secure' the greatest area that was logistically feasible, with the proviso that the financial core around the Bank of England was included. The police were also conscious that a great deal of manpower would be needed to guard the checkpoints, thus by setting up the cordon as they did the police could minimise the number of entry points into the secure zone, and hence the personnel needed to run the scheme effectively. The benefits of the new cordon were demonstrated later that month when police sources believe that the Provisional IRA attempted to penetrate this cordon by sending in a bomber on public transport hence avoiding the possibility of vehicle checks. The bomber was arrested and found to be carrying an 8lb-semtex bomb moulded around a petrol-filled milk bottle (see for example Webb, 1996).

The Spatial Impacts of the Cordon

As previously noted, certain businesses located on the edges of the Square Mile, outside the ring of steel, were concerned that bombs could be planted at the edge of the newly constructed cordon demonstrating the ineffectiveness of the Police's strategies in the City. The City Police were well aware of this Provisional IRA tactic from experiences in Belfast. As a result they set up a so-called 'collar-zone' around the ring of steel, which saw increased police patrols and roving checkpoints. The activation of a collar zone was made possible by additional staff attached to these units. The 1993 Annual report of the Commissioner of the City of London Police (1994) drew attention to the initial establishment of the experimental cordon, and the funding for 57 extra police constables provided by the City so that armed checks on vehicles could be increased. Overall in 1993, of the 3560 roadchecks conducted in England and Wales, over 3200 were conducted by the City Police. This resulted in 5867 vehicles being stopped. As a result 168 people were arrested, but *none* for terrorist offences.[26]

Michael Cassidy indicated that he felt that the collar zone was increasingly vulnerable:

> It's just common sense that they were going to be exposed bearing in mind that there were some very senior businesses who were unfortunately just outside the cordon.

However, the police highlighted that by setting up the collar zone they attempted to treat the risk level in this area realistically. Indeed, the City Police did not see the collar zone as being at greater risk from attack:

> This area is not at any higher risk because in complementing the inner cordon we have armed patrols and random road-blocks and all the other things in place which support it.
> (Senior police officer)

In short, the collar zone was a concerted attempt to alleviate the fears of those businesses within this area:

> There are certain targets within the City that receive increased security attention, but the general philosophy of the police was to spread resources right across the City. Indeed, when the threats were perceived as high the police actively increased foot and vehicle patrols outside the cordon to redress the balance, as there were complaints from those outside who had the same exposure to risk but were receiving less protection.
> (City security advisor)

Indeed, at the Police Committee an elected member questioned the amount of policing available to those parts of the City outside the ring of steel.[27] The Commissioner of Police in reply acknowledged that the local policing plan prioritised anti-terrorism activities in the secure zone given that this was believed to be the most likely target.

In short, the heightened fear of terrorism not only affected the area of the City outside the ring of steel but also impacted at other spatial scales. First, at a localised level where complaints were received from those in neighbouring boroughs that the ring of steel could serve to export the risk of terrorism (and traffic) to their area; second, by increasingly fortifying the City, it was argued, by some, that this could redistribute the physical risk of attack to other key targets and locations, especially those in London. It was not until after the bomb in the London Docklands in 1996 that this fear was fully realised.

The agencies of security in the City, in this sense, acted in a very paternalistic way by protecting the majority of their territory and leaving others to look after their own areas. As one security advisor working in the City noted, 'our job is to look after number one, the Square Mile. What happens outside our boundary is not our concern as we have no jurisdiction outside the City'. Furthermore, in the months after the ring of steel was set up, it was also noted that plans to extend the cordon might be necessary in the light of the 1994 local authority boundary changes, after which the Broadgate centre to the north-east of the City was to come under the Corporation's jurisdiction. This represented a very important economic asset that need protecting.[28]

Creating an Electronic Panopticon

After Bishopsgate, in addition to access restrictions the City began to enhance its electronic surveillance capabilities, and aimed to develop three separate but interdependent camera systems to deter terrorism. First, a modified ATC camera system; second, a series of police cameras at entry and exit points; and third, by enhancing the number, and overall coverage, of private CCTV systems. The development of CCTV in the City occurred for a number of reasons:

> CCTV has two implications related to the IRA. A principal aim of a IRA terrorist when carrying out an attack is not to get caught, therefore CCTV is a highly visible deterrent both when carrying out an attack and when planning one through surveillance. The Bishopsgate bombers, for example, were caught on film.
>
> (City security advisor)

He continued:

> The key aim of CCTV in this respect was to alter the perception of risk for the IRA in the hope that they will go elsewhere and export the threat. The message CCTV gives is – yes you could get away with planting a bomb, as this is very difficult to stop, but you could well get caught on film which will be circulated all over the world including Ireland and therefore you will always be looking over your shoulder.

The ATC camera network had been in operation for a number of years and was used primarily for management of traffic. It was first updated in 1991 with the addition of some colour cameras and further modified after the 1992 blast. Following the Bishopsgate bomb it was suggested that this camera network should be extended by the addition of eleven colour cameras to provide coverage on minor city roads.[29]

In addition to traffic management CCTV, police operated security cameras were erected to monitor the entrances into the Square Mile, 24 hours a day, so that every vehicle entering the security cordon was recorded. From November 1993 there were two (or more) cameras filming at each entry point – one that recorded the number plate, and the other that scanned the front profile of the driver and passenger (Figure 5.5). These provided high-resolution pictures, which could be compared against intelligence reports and photographs. The Commissioner of Police indicated that almost nine months after this camera technology was first used it had been exceptionally successful and was attracting a good deal of attention from elsewhere.[30] By the end of 1993 there were more than seventy police controlled cameras covering the City (ATC and Entry and Exit CCTV), which aimed to create an environment where would-be terrorists could not hide. Cameras are seen as highly symbolic 'electronic guardian angels' (Davis, 1990) and were put up, despite complaints from civil liberty groups on the basis of the police rhetoric of 'if you have done nothing wrong you have nothing to worry about'.

However, there was still inadequate coverage of many public areas due to lack of private cameras. It was further noted that concerns over too many private cameras operating in the City meant that they initially attempted to use those they already had more efficiently:

> At this time there was a perception that the City of London had a tremendous number of cameras in place and the view was taken that rather than try to impose another regime of cameras it would be better to make best use of the cameras that were already in place.
>
> (Senior police officer)

He continued:

> We conducted a survey [in 1993] to find out what the extent of the coverage was and much to everyone's amazement there were only something like 160 cameras that offered coverage of public areas.

As a direct result, the police launched an innovative scheme called CameraWatch in September 1993, to co-ordinate camera surveillance and create an effective, and highly visible, camera network for the City.[31] CameraWatch aimed to deter terrorism and other criminal activity by the knowledge that CCTV systems were operating, and to provide a means of detection and evidence gathering should offences occur. The scheme was also seen as a potential catalyst for other security and community safety initiatives, the fostering of close co-operation between organisations helping them to achieve mutually desirable security goals, and the development of institutional links between City organisations, the City Police and the Corporation of London:

> CameraWatch aimed to get the City involved by setting up cameras on the side of buildings as a visual deterrent. It tried to get the City to talk amongst itself and to create an office-watch system involving teamwork and co-ordination between a number of businesses with the rich subsidising the poor...
>
> (City security advisor)

CameraWatch in essence aimed to be proactive to the terrorist threat. It aimed to protect 'areas' and not just rely on the occupiers of individual buildings to protect themselves, given the realisation that modern bombs affect huge areas and not just the target building. As such there was little point of individual buildings developing their own independent surveillance systems in isolation.

However, the installation of cameras on City properties, as part of this scheme, was not always a feasible option due to leasing and heritage considerations:

> Many businesses are only tenants in the buildings and were reluctant to pay-out for expensive surveillance equipment. Additionally there were objections from conservation groups about the aesthetic intrusion caused by cameras, which have to be overtly visible. Therefore it was not just a matter of putting cameras up.
>
> (City security advisor)

This second point was reinforced just before the Bishopsgate blast when the Corporation's Planning Department issued *Planning Advice Note 1* in April 1993 (subsequently revised in November 1997), which dealt with design considerations of CCTV. This indicated the need for such cameras given the terrorist threat, but that environmental consideration should be given to their placement and size:

> Security Cameras have now become necessary in many cases, but their insensitive siting, and sometimes the size of the installation, can be visually detrimental to buildings and their surroundings.

The advice note continued:

> Where cameras are proposed for general street surveillance, applicants will be encouraged to discuss their proposals with the City Police as well as adjoining frontages in order to achieve a mutually agreed system and thereby avoid the unnecessary proliferation in the street of several cameras performing the same task. An excessive form of street surveillance by a number of cameras on prominent display is something the Department of Planning would want to discourage, as it would drastically affect the character and general ambience of City streets.

City businesses were once again encouraged to liaise with the police to see if reciprocal arrangements with nearby businesses could be entered into, and to use the smallest cameras appropriate. These were to be installed as high on the buildings as practical, and where possible to paint the camera unit and cable the same colour as the building. If the occupiers of a listed building, of which there are many in the Square Mile, wanted to put up cameras then special consent was required. This planning advice, although based on heritage considerations, contradicted the advice of security advisors who wanted cameras to be as overt as possible to actively deter criminal activity. Small concealed cameras would in this instance be seen as less effective in achieving security goals.

Nine months after CameraWatch was launched, still only 12.5% of buildings had camera systems leaving a very large proportion of public areas without the security of constant CCTV coverage. However, by 1996 over 1000 private security cameras (in over 376 camera systems) were operational in the City, due in large part to the efforts of the police and Corporation security advisors in encouraging those who had previously not installed CCTV to do so in liaison with others. It was also suggested that in the future it might be possible for these private systems to be linked into the police camera systems when the need arose (Kelly, 1994a).

The panopticon of surveillance that was constructed after Bishopsgate served to reinforce the access restrictions that were imposed. Through these territorial strategies, space within the Square Mile was increasingly organised in such a way that almost complete surveillance coverage became possible, thus reducing the perceived opportunity for terrorist attack. The aim of the police was to create a landscape of cameras which both monitored and controlled space but also actively deterred criminal activity because of their symbolic function.

Camera technology was perhaps the single most important factor in the City Police's counter-terrorist campaign. This was highlighted during the trial of eight suspected Provisional IRA terrorists in London in 1997 who had been involved in a plot to blow up various electricity sub-stations in and around London in the early months of 1996. At the subsequent trial one of the accused gave the immobilisation of the ring of steel as one of the key aims of this attack: 'If the IRA were capable of closing down all electricity in London without going into London, it would make the ring of steel null and void.'[32] By implication this meant that the Provisional IRA could then have more easily planted a bomb in the City without being detected, as it would have taken hours if not days to restore power to many parts of London.

Fortifying Individual Buildings

As well as developing area-based security strategies, the police, in liaison with the private security sector, set about creating an extra layer of security around individual buildings through the addition of landscape features and the implementation of security strategies. After the first bombing, enhanced security measures were not particularly common in the City as many businesses saw the St. Mary Axe blast as a one off attack. However, the increasing way in which the police and private security staff were collaborating was amply illustrated in June 1992 when security guards on a routine patrol of their building discovered a bomb under a car in Coleman Street. As such, the private security industry was co-opted to help the anti-terrorism effort. Their tasks commonly included being involved with contingency planning, exterior patrolling of premises, monitoring of the areas outside their building with CCTV, and promoting crime and terrorism prevention within their company.

Such approaches were increasingly undertaken in the City following the 1993 bomb for a variety of reasons. First, certain businesses because of their prominence or nationality, felt vulnerable against attack, not just from the Provisional IRA. For example, according to security advisors, the French banks felt at risk from Algerian terrorists, and Turkish businesses from Kurdish extremists. The second reason why individual buildings were fortified was because they perceived that they were next to a vulnerable target such as the Bank of England, the Lloyds Building or the Stock Exchange. A third reason private buildings improved security was due to a possible insurance discount that they could achieve for introducing risk-management measures (see Chapter 6).

Typical security plans adopted at this time aimed to deter an attack and deny the terrorist access to the premises – creating what could be referred to as a 'mini ring of steel' – which served to fortify and 'harden' the urban landscape by altering the spatial configuration of the outside area adjacent to a building, in particular the borders of a building and the immediate surrounds. It aimed to do this by defining an area of control and delimiting a building from its surroundings in a number of different ways: first, businesses attempted to control access to their building through limiting the number of entrances, and through control of visitors by security personnel in the reception area; second, by restricting access to areas under the building such as car parks and storage-areas; third, through an increase in surveillance through organised means via vigilant security guards, and by CCTV; and fourth, by creating no-parking zones around the building to stop vehicle bombs being parked close-by.[33]

The City Police and the Corporation of London, as well as various insurance bodies and the Loss Prevention Council, all gave advice to businesses about what they could do to their premises to minimise risk of attack highlighting the increased number of risk management institutions who were concerned about the impact of this risk. In particular, the Corporation was increasingly proactive in getting individual buildings fortified, by employing specialist security advisors to liaise with business. As Michael Cassidy commented:

> One of the things I did after Bishopsgate was to get a security specialist in to advise business how they could protect themselves for example by joining CameraWatch and getting better contingency plans.

The security individual buildings can also be seen in terms of territorial reinforcement, which had the effect of adding an additional complementary layer of physical defence to the City. The subsequent retrofitting of crime prevention measures to 'design out terrorism' led to alteration in the shape and function of a number of streets and public places within the city. Others strategies, however, took the form of subtler landscape changes less associated with counter-terrorism such as the use of bollards to restrict vehicular access near their building. There were also attempts at clear demarcation of private areas to define territorial boundaries – creating an exclusion zone around buildings with the external configuration of many buildings in the Square Mile effectively being target hardened.

The Overall Effect of Post-Bishopsgate Fortification

Attempts to deter further terrorist attack after the 1993 bomb led the police and other agencies of security to attempt to 'harden' the landscape at a number of spatial scales within the City. First, through the construction of a ring of steel, and second, through specific target hardening undertaken in liaison with the private security industry. Together this formed what a senior security advisor referred to as an 'onion skin effect' – essentially a series of security layers which potential terrorists must pass through in order to reach the desired target. The attempts of the agents of security amount to the privatisation of space – the creation of a territorial container, which attempted to hermetically seal the core of the City from its geographical neighbours and selective parts of the Square Mile, deemed less economically important. Such localised boundary formation and reinforcement was however seen as vital if the City was to maintain its position in the global economy.

Figure 5.6 Police warnings of the risk of terrorism

After the Bishopsgate bombing the City police also set up a Pager Alert system which contacted all users of the system simultaneously and delivered a message within thirty seconds to provide information on security alerts. The majority of the businesses in the Square Mile (and later Canary Wharf) signed up for the scheme which allowed them to react immediately to any alerts received.

> Protecting lives is our number one priority in any major incident. The pager alert scheme saes valuable minutes, even seconds, in communicating that there is a security threat.
>
> (City of London Police news release 2001)

In addition 'warning' signs were also put up around this border indicating the dangers of terrorism and the need to stay alert (see Figure 5.6). Importantly however, many aspects of this territorialisation moved away from the popular 'conflict' approach, which sees the urban landscape fragmented and divided into mutually hostile units. Instead, far more symbolic and subtle notions of territoriality were displayed, which expressed a shared concern with defence but which attempted to balance the need to security with the continual functioning of the City as a business centre.

Optimism

Immediately after a Provisional IRA ceasefire was called on 31 August 1994 there were suggestions, in the media, and from some businesses in the City, that the ring of steel should be scaled down as it was seen as unnecessary given that the perceived risk level was reduced. Such suggestions were heeded to an extent as permanent armed guards were taken off most of the checkpoints, producing a less overt police presence on the street. This was a very visible indication that the risk of terrorist attack had decreased in the eyes of the Police. This was the only operational change that the City Police introduced at this time. As Michael Cassidy commented:

> There was absolutely no call from our people for stepping down the security...the only thing that happened was that operationally the Commissioner of Police used to rotate his manning of the entry points on a completely random basis so that sometimes they were unmanned.

This process was summed up when during and interview conducted during the cease-fire period:

> It [the ring of steel] was highly visible – an in your face deterrent with armed check-points and a highly visible police presence. Today the threat profile has been reduced and these measures have been scaled down.
>
> (City security advisor)

However, he further noted that the ring of steel was in a state of temporary suspension:

> The ring of steel's function as a security initiative will be maintained. It will provide a framework in which to launch a security operation if needed.

This downgrading of security did however have a noticeable influence on recorded crime levels. It was reported that during the ceasefire recorded crime in the City had increased and that this pattern was being monitored to assess whether this was related to the recent changes in high profile policing.[34] For example, on 29 November 1995 it was reported at the Police Committee that during the previous six months a noticeable increase in crime had occurred despite the fact that the City was still viewed as a high priority target by international terrorist groups. Despite downgrading security the police still acknowledged that:

> During the last six months, the continuation of the Northern Ireland peace process has been heartening, but the entirety of measures put in place within the City were not wholly in response to PIRA terrorist threats.

The report continues:

> Nevertheless, in the past year, some changes to the high profile policing measures introduced after the last PIRA bombing of the City have been inevitable. It is noticeable that levels of recorded crime have been rising since the measures were changed despite the number of alternative initiatives introduced.

Figure 5.7 shows the significant reductions in crime that occurred in the City during the period from 1990 until the end of 1995. In particular it shows how crime fell drastically after both major bombing incidents in April 1992 and 1993 as a result of a high profile approach to City policing. However, it also shows a gradual increase in crime during the ceasefire period from August 1994 onwards until the end of 1995 coinciding with reductions in high profile policing.[35]

In line with the ceasefire, from August 1994 permanent bollards, paving, and in some cases flower beds, began replacing temporary traffic cones (see Figures 5.8, 5.9) creating a less visible form of landscape alteration which was trying to remove overt security referents. In addition, the Commissioner of the City Police indicated that over the next few years the street environment of the City around the security points would be 'landscaped to give the scheme an aesthetic permanence in keeping with City street architecture' (Kelly, 1994a).

In addition, the scaling down of security was demonstrated in early 1996 when a decision was made to re-install around 40 litterbins in the Square Mile. This decision was suspended after the Docklands blast on 9 February.

The aforementioned security changes that occurred as a result of the ceasefire demonstrated a noticeable 'softening' of the landscape as a result of reduced threat levels. These moves all indicated that the City, at this time, felt more relaxed about the risk of attack. This however made it increasingly difficult for the police to call on public support to help them combat the threat:

> The business community were also beginning to feel that the risk of further terrorist attack had gone away, making it more difficult for the police to sustain a campaign of raising or maintaining public awareness of the need to be vigilant.
>
> (Senior Police officer)

In particular, there was an attempt on behalf of a number of prominent organisations, around Christmas 1995, to persuade the Corporation to disband the security cordon altogether. The situation going into 1996 was, however, one of optimism that the cessation of violence would continue, although the police were still being realistic about the potential threat, not just from the Provisional IRA:

> For a number of years, policing in the City [has been] significantly influenced by the necessity to respond effectively to the criminality of Irish Republican terrorists. During the past year, the cessation of this criminal terrorism has been heartening and optimism for a lasting settlement remains high. However, the measures put in place and being operated in the City are to deter all terrorist criminality. The City will always be a potential target for acts of terrorist criminality because of its high profile, high economic value and cosmopolitan business community.
>
> (Corporation Police Committee minutes, 31 January 1996)

The police camera network had been continually upgraded to meet this terrorist threat. In the early months of 1995, the new high-resolution cameras for the traffic system were installed and 13 further cameras were added to monitor the cars exiting the City.[36] Before this date security cameras only focused on cars entering the cordon. Exit cameras were particularly important, as police could now monitor traffic into the City and, if needed, track suspect vehicles across the City.

Furthermore, during the ceasefire, plans to extend the ring of steel westwards were being drawn up to cover a greater proportion of the City in the secure zone (this was first suggested in 1993 so that the Broadgate centre could be included):

> It is now envisaged that the size of the ring can be increased westward to cover areas such as St. Paul's. This is out for consultation at present and is being dealt with by the Transport Department. It is seen as an early plan. It represents a philosophy in relation to pollution, traffic and crime.
>
> (City security advisor)

This meant that once again the City attempted to justify the ring of steel on grounds other than counter-terrorism (see also Chapter 7). After the Bishopsgate bomb the police justified references to the rings' counter terrorist function as the risk level was perceived to be high. During the ceasefire period, however, overt references to the risk of further Provisional IRA attack were, perhaps, not politically acceptable, and hence improvements in transport, environment indicators and crime figures were used to justify the merits of the ring of steel and to push for its expansion.

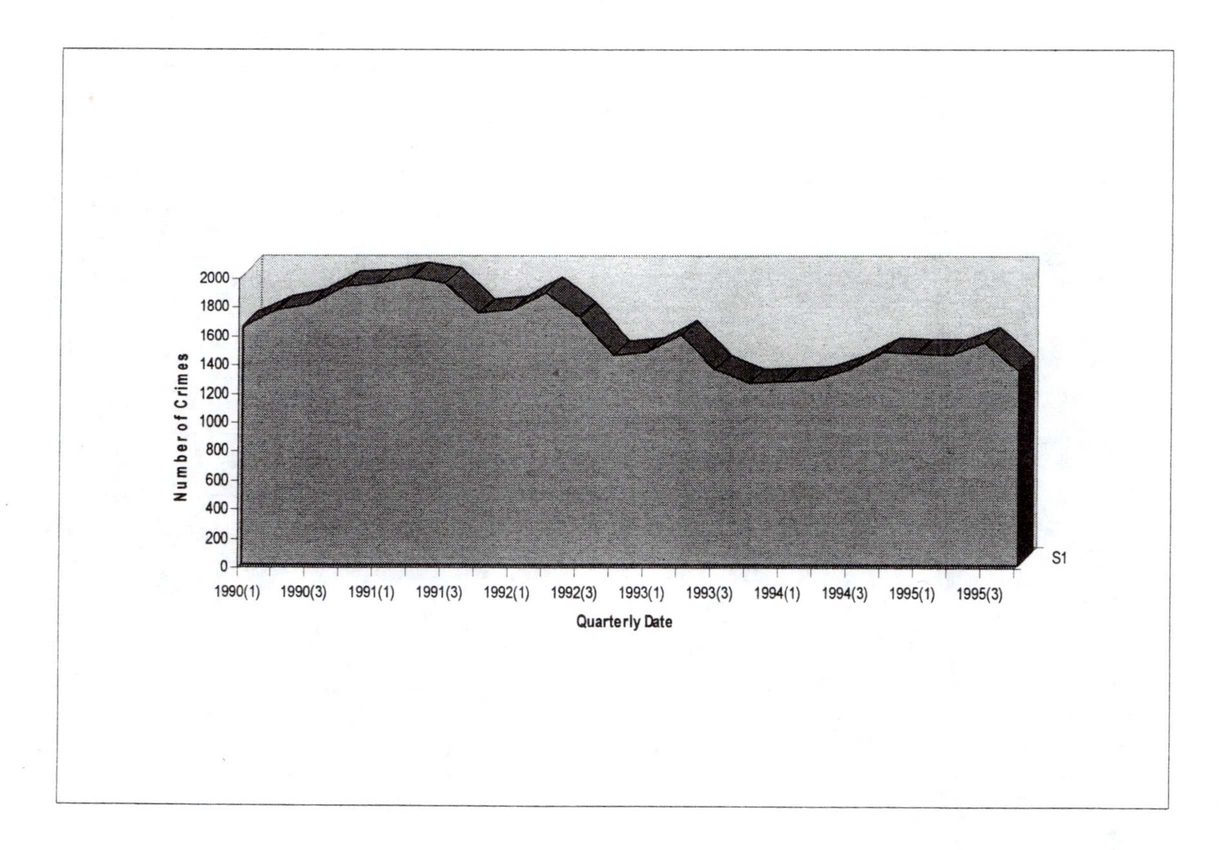

Figure 5.7 Recorded crime in the City of London 1990-1995

Figure 5.8 **Example of how the ring of plastic has become 'concretised' as permanent bollards have replaced traffic cones (I)**

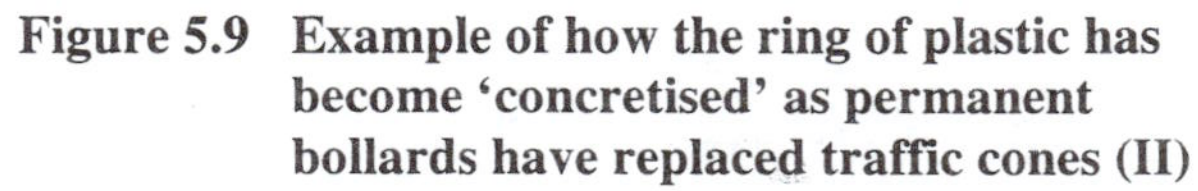

Figure 5.9 **Example of how the ring of plastic has become 'concretised' as permanent bollards have replaced traffic cones (II)**

Reactivating and Extending the Ring of Steel

The Provisional IRA's offensive strategy, as previously stated, was temporarily halted between August 1994 and February 1996 as a cessation of military activities was declared. In time, what Sinn Fein, the Provisional IRA's political wing, saw as a lack of movement on this front led to the breakdown of political dialogue and the return to the twin-track approach of the bomb and the ballot box. However, as Myers (1996) ironically commented, just a few days before the cessation ended, in relation to a spate of so-called punishment beatings on Catholics suspected of anti-social behaviour such as drug-pushing and joyriding (believed by the RUC to be sanctioned by the Provisional IRA):

> The cease-fire is over, but only selectively so – on a sort of deferment plan. An equivalent butchery of Northern Catholics by British Soldiers would by now be causing bombs to detonate in the CITY OF LONDON where the war will most likely resume, if it ever does [emphasis added].

The very overt security presence at the entrances to the City, that had become an almost permanent landscape feature of the Square Mile, decreased as the risk of further attack declined during the cease-fire. This cessation of terrorist violence lasted until early 1996 when a large 500lb bomb exploded at South Quay in the London Docklands, seen as a symbolic extension of the Square Mile.

Following this bomb the fortress mentality returned to the City and the full pre-cease-fire ring of steel was reactivated and operational within a number of hours as there were fears that the City would be attacked:

> The resumption of IRA terrorist criminality earlier this year has once again highlighted the need for the City of London to be ever vigilant. The traffic control points are now equipped with entry and exit closed-circuit television cameras, and officers are deployed at the entry points, as judged necessary in response to the criminal threat.
>
> (Corporation Police Committee, minutes 29 May 1996)

Initially, as a result of the increased threat level there was a large increase in a high visibility policing at both the entry points and on the City streets in general. There was also an increased frequency of roving checkpoints.

After the Docklands bomb, further proposals to increase security in the Square Mile were made. Such suggestions centred on a proposed westward extension of the ring of steel that was initially put forward in February 1995. This proposal was in line with the Corporation of London's Unitary Development Plan but many newspaper reports in the days following the Docklands bomb inaccurately portrayed the possible extension of the City's security cordon as a new idea. In particular, it was felt the extension would discourage through traffic from using local roads, reduce the conflict between pedestrians and traffic on local roads, improve the environment in a larger part of the City, and give the Commissioner of Police improved security opportunities on City streets, within the larger zone. It was thought that the proposed extension would cover nearly 75% of the Square Mile and would divert an additional 10,000 vehicles a day from the centre of the City. Some however thought that this would create traffic gridlock.[37]

In April 1996 the Corporation's Policy and Resources committee agreed to debate these ideas. On 23 July the Planning and Transportation committee agreed to support these proposals, an idea that was welcomed by the City of London Police and the Policy and Resources Committees (17 October). On 26 November the Common Council and the Planning and Transportation committee further debated the merits of the extension. Feasibility studies involving computer modelling of traffic flows had by this time been undertaken, in particular with regard to getting St Paul's Cathedral, Smithfield and the Barbican areas within the security zone. At the Policy and Resources meeting (17 October), however, concern was expressed by some members about the need for adequate consultation (see Chapter 7) and the possible restriction of business traffic to the City. In addition to the extension of the security cordon the City Police also set up more 'Red Routes', which are common in other part of London. In these areas parking is strictly prohibited at all times. The aim of such routes is to improve the ease of traffic movement, whilst also benefiting the counter-terrorist operation.

Some commentators also saw the 're-steeling' of the Square Mile as an opportunity to improve the environment in the City:

> The ring of steel thrown up around the City as a defence against terrorist attack shows that it is possible to make radical shifts in traffic without the capital grinding to a halt. Indeed, with security measures again being strengthened there is an excellent opportunity to pedestrianise many streets permanently.[38]

This once again drew on Belfast experiences where the shopping and commercial area is sealed off from most traffic by steel gates and anti-crash bollards. Potential pedestrianisation in the City was therefore seen to have two benefits. First, to help improve the environment by reducing vehicular numbers, and second, to reduce the manpower the police needed to use as there would be fewer roads to monitor for potential vehicle bombs. The Commissioner of the City's Police (1996) was in strong support of the proposals:

> I believe an extended zone will be of considerable benefit to the traffic and environmental conditions in the City... A by-product of an enhanced traffic zone would be the opportunity to introduce security measures (as necessary) in a manner similar to that currently attached to the present traffic zone. The threat to the City from criminal terrorism is high and whilst I judge that will change from time to time the source of the threat now and in the future is not one-dimensional and a strategic approach is essential.

Despite the police downplaying the anti-terrorist importance of the ring of steel, the extension, could, at this time, be seen as primarily an attempt to keep the bombers out of the Square Mile, given that the City was still on a state of high alert about the possibility of further attacks by the Provisional IRA.

The proposal for a western extension was ratified by the Court of Common Council on 5 December 1996 and implemented on 25 January 1997, becoming operational two days later for an initial period of six months under Section 12 of the

Road Traffic Regulation Act 1984 (see Figure 5.10). This figure shows the new check-points that were added along Queen Victoria Street, Ludgate Hill, Holborn Viaduct and Aldersgate Street. A number of minor roads were also been blocked off as access into the City had been re-arranged with prominent places such as St. Paul's included. Despite fears of traffic gridlock, initial indications contained in an evaluation report and private correspondence received from the Corporation on the extension suggested that there had been no significant problems caused by the modifications.

On 3 February 1997, just after the extension was put in, the new cameras at police checkpoints were linked, first to the Police National Computer (PNC), and then to the vehicle database and 'Hot List' of vehicles at the Force Intelligence Bureau. These cameras were capable of zooming in and out and swivelling through 360 degrees and were fitted with lights, which enabled round the clock monitoring. The digital number plate recording technology they used could process the information and give a warning to the operator within four seconds. This considerably increased the capability of the City of London Police to run vehicle checks. Such technology was developed from experiences in the first Gulf War in 1991. The project to install this advanced system began in early 1995 with installation commencing in September 1996. This was completed in January 1997 with the system going on-line in February 1997. This coincided with the opening of the western extension to the ring of steel.

It is also important to note that from a legal perspective, in April 1996 the ring of steel was given another layer of protection by new legislation relating to the Prevention of Terrorism Act, which allowed the police to search pedestrians randomly, as well as cars, in and around the Square Mile.

Fears of incendiary devices being smuggled into the City had long been a concern of the City of London Police. Signs were initially put up in the City warning visitors that they were liable to be searched. This it was felt was an invaluable tool, as once again, lessons from Belfast had indicated that when confronted with a security cordon the terrorists had a tendency to switch tactics and attempt to smuggle hand held explosive devices into the desired target area. As a Michael Cassidy stated at this time:

> It's a very useful extension of powers and it plugs a gap from our point of view because clearly the inability to search inside someone's big coat, which was the law beforehand meant that semtex could be walked into the City.

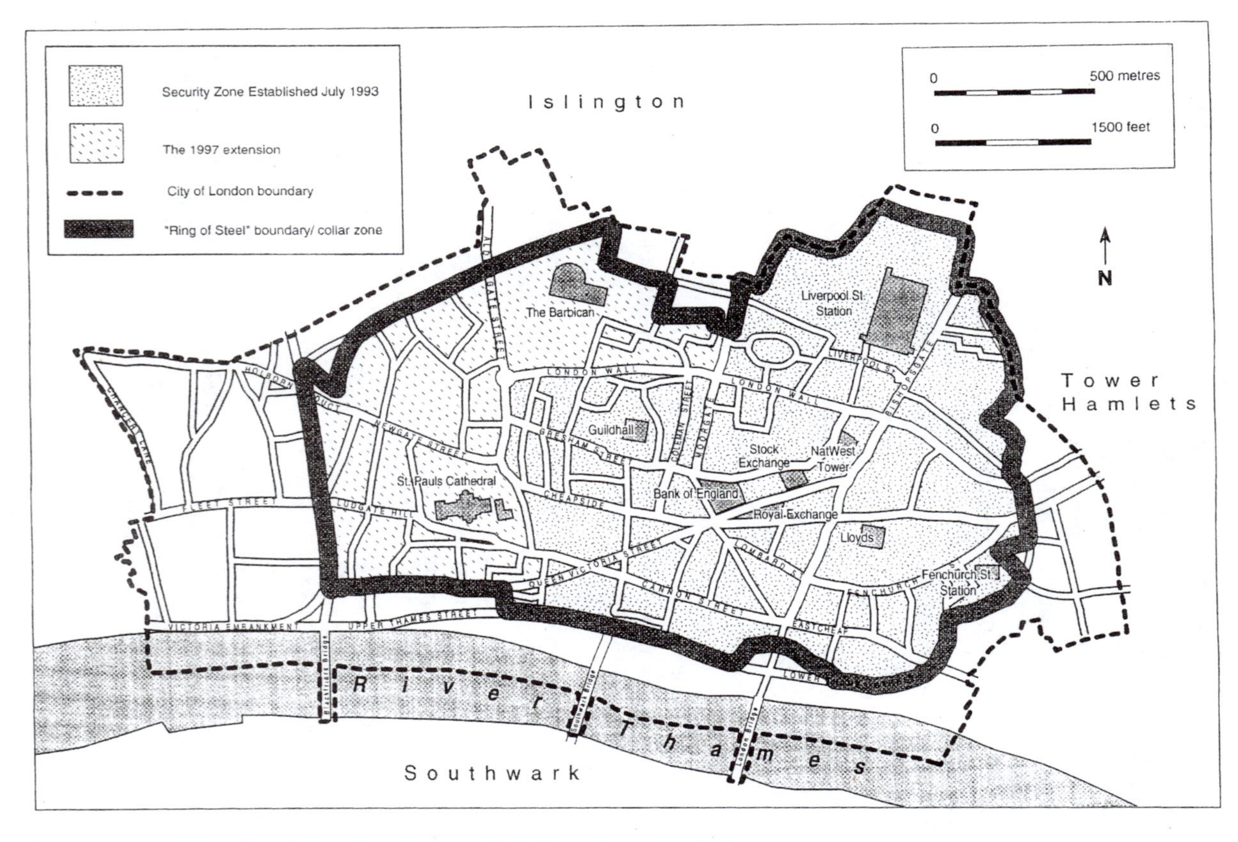

Figure 5.10 The City of London's extended ring of steel (1997)

Re-appropriation

Towards the end of the 1990s, with paramilitary ceasefires in place in Northern Ireland and a reduced state of terrorist alert in the Square Mile, the full ring of steel became relatively dormant. However, the City, like a number of other strategic global sites came under threat from anti-capitalist protesters. Initially a large scale demonstration was to have been held in the City in May 1998 but access restrictions and the blanket CCTV coverage given by the ring of steel, and, the fact that the planned event was going to be held during the weekend (meaning the City would be empty of office workers) meant the event was postponed (Do or Die, 1999).

A year later on Friday 18 June 1999, this time a workday in the Square Mile, between 6-10,000 demonstrators under the collective banner of J18 assembled in the City for a worldwide 'Carnival against Capitalism' to coincide with a G8 economic summit in Cologne. This led to a massive mobilisation of Police drawn from the City as well as the British Transport and Metropolitan forces. As a result of subsequent rioting, damage was put at £2 million with many landmark buildings attacked and over forty people injured. As emotively noted by the Lord Mayor of London (cited in *The Sunday Times*, 20 June 1999):

> It was wanton terrorism, anarchy. The riot police came to rescue and impossible situation. That was nothing short of a war zone.

During the demonstrations the ring of steel entry points were manned by Police who monitored the flow of people into the Square Mile. Although the anti-terrorist cordon could do little in preventative and legal terms to stop the vast number of people entering the City it could use its extensive camera systems to monitor the event and subsequently pinpoint those involved in the worst of the violence. To an extent this tactic was nullified as a number of protestors who immobilised cameras by spray painting the lenses or covering them with plastic bags. In the aftermath of the disturbances the City Police published a large gallery of CCTV photographs of those they wanted to identify in relation to the rioting. It was reported that over sixty officers were working full time to look at over 5000 hours of CCTV coverage and other evidence in order to make arrests (Do or Die, 1999).

However, some initiatives put in place to counter the terrorist threat were seen as invaluable to the police before, during and after the riots such as the Pager Alert system which was used to convey updated messages to subscribers. Furthermore, City security advisers suggested to businesses that they increase security around their buildings well in advance of the planned demonstration.

As a result of the J18 riot and subsequent smaller demonstrations, much closer liaison occurred between the three police forces involved and new protocols were established, centred on zero tolerance that were to be used for future demonstrations.[39] This was aided by the new Terrorism Act (2000), which came into force in February 2001 and broadened the scope of what could be considered to be terrorism, meaning activist movements could potentially be labelled as such especially where there are allegations (as with the JI8 disturbances in the City) that the violence was pre-planned.

Despite the focus on thwarting anti-capitalist demonstrators, counter terrorism was still at the forefront of the City Police's thoughts. During 1998, 1999 and into the new millennium the terrorist threat was still being seriously considered. As the Commissioner of Police noted in early 2000: 'Terrorism, thankfully, has not affected us recently and only time, coupled with our collective vigilance, will tell if that is to remain the case' (*City Security*, 4/2000, p.3). During this period, in line with the Crime and Disorder Act 1998, the City Police rolled out strategies which increasingly attempted to generate partnership between themselves and the business community to reduce crime and terrorism, updated and problems with the Pager Alert schemes has led to the introduction of an e-mail alert system. However in the City the crime rate did rise during 2000 (8066 offences compared to 7577 in 1999), mainly through theft. This reflected a similar rise that occurred in the 1990s when the terrorist threat was considered less severe and police presence on the streets reduced.

During 2000 and 2001 the threat against the City of London was again increased by the terrorist bombing campaign of dissident Irish republican groups, the Real IRA[40] who were responsible for bomb or rocket grenade attacks against a number of prominent landmark buildings in central London such as the BBC and MI6. As a senior City Police officer indicated the Square Mile was a risk:

> There is a clear and credible threat from Dissident Irish Republican Terrorists particularly the Real IRA… The Greater London area will always be seen as the most attractive target as it will raise their profile in the media. We all need to be vigilant at this time and remember that they pose a HIGH threat to the UK mainland.[41]

At this time further permanence was given to certain sections of the ring of steel. An extension was put in place around the Broadgate area in the north east of the City and came into effect on 9 April 2000.[42] The City Police were in favour of such an extension. The Commissioner of Police noted that:[43]

> This is an effective traffic management scheme which offers a security opportunity that will enhance community safety for a significant area of the City and create benefits for the business community, residents and visitors. The scheme and the policies which underpin it are sound and have our fullest support.

The Metropolitan Police were also consulted by Islington Borough Council but and had no objections to the proposed scheme.

As with the 1993 and 1997 versions of the ring of steel, this scheme was referred to as an experimental traffic management scheme, pending consultation and reports as to its effectiveness. Discussions for example occurred between the Corporation of London and the London Boroughs of Tower Hamlets, Islington and Hackney where the knock-on effects of traffic displacement could have occurred. Three months after its introduction 'no serious problems' were noted.[44] On 16 October 2000 it was agreed by the Corporation of London to recommend that the scheme be made permanent subject to the agreement of the surrounding boroughs and Transport for London. This process was finally completed in March 2001.[45] After this extension had been made permanent ANPR camera technology was added in a similar way to other areas of the ring of steel. The development of this part of the ring of steel will be further discussed in subsequent chapters.

Re-appraisal

The unprecedented events of 9/11 led to an immediate counter response from the City of London Police. Just as with the immediate aftermath of the Docklands bomb in 1996, the ring of steel swung back into full-scale operation. This was part of a co-ordinated London wide operation which saw over 1500 extra Police patrolling the streets of the capital. In the Square Mile the Police also undertook special liaison with American firms to improve their security through extra patrols as well as instigating a far greater number of stop and search checkpoints. The Corporation of London also re-examined its own emergency procedures through collaboration with senior businesses and security professionals, as well as recommending to all business that they reassess their contingency plans with the help of City Police (Mayhew, 2001). The initial approach adopted, drawing on the previous experiences of terrorist attack, was very much 'business as usual' and 'vigilant but calm'. As a Corporation of London Press release (12 September 2001) stated:

> The City is carrying on with business as usual. The City of London has had robust security measures in place for many years to deal with any terrorist threat and these are in operation now, as they are 365 days a year. We have been in contact with many businesses across the City and our message to them has been to remain calm, be vigilant and ensure that their own contingency plans are in place. These security arrangements are regularly reviewed and will again be examined in the light of yesterday's events.

As further noted by a senior Police officer:

> The various communities of the City of London have had to re-appraise their approach to the threat posed by terrorism... those who live in the Square Mile have... had to live with the threat of terrorist attack for more than three decades. The positive aspect of that experience is that it makes us uniquely prepared to confront the new threats posed by global terrorism.[46]

The subsequent response involved not just the City Police but Scotland Yard's anti-terrorist branch and the Metropolitan Police who jointly reviewed and reappraised the counter terrorist strategy in place around the Square Mile. In the aftermath of 9/11 subscriptions to the City Police's Pager Alert and E-Alert emergency communication systems increased by 44% and 139% respectively (City of London Police, 2002). In addition, the success of the City Police's Pager Alert in effectively sending out early warning messages to businesses in the light of 9/11 has meant the scheme is now being rolled out on a London-wide basis with the help of the Metropolitan Police,[47] and potentially to other UK cities in the future. Plans to link these systems to the mobile phone network are also underway. As a leading

Metropolitan police officer cited in a City of London Police new release in November 2001 noted:

> This is a time of heightened tension and schemes like pager alert can help businesses respond quickly to security alerts and other incidents. The system works extremely well in the capital and in time other UK cities would benefit from a similar scheme.

The City Police also moved quickly to relay messages that they were prepared for attacks from both conventional weapons as well as unconventional biological and chemical agents although for the latter the risk was considered slight.[48] Echoing similar messages relayed in the aftermath of attacks against the City in the 1990s, public vigilance was seen as 'the most formidable weapon we can deploy against terrorism' (ibid.). In short, balancing security needs with realistic threat assessments were seen as paramount. As one senior source highlighted:

> There is a debate between some people who think we should throw everything at guarding buildings and other who want to respond to a specific threat... At the moment there is no specific threat. But there was no intelligence before the World Trade Center, so do we assume the worst and expect the possibility of suicide bombers and threw everything and protecting London now or do we react when there is intelligence. It is a dilemma.[49]

9/11 refocused the City Police's minds on counter-terrorism, especially potential attacks by dissident Irish republican groups. However, 9/11 also added a new dimension to defending the City as potential perpetrators are operationally different. In short, as the Commissioner of Police noted, some terrorists are unlikely 'to be deterred by the high levels of technical surveillance we have successfully used against domestic terrorists who seek to avoid identification, arrest and prosecution as part of their operating methods' (City of London Police, 2002a).

Although the international terrorist threat was considered high in the year proceeding 9/11, the City quickly returned to business as normal, although security was conspicuously on a higher state of alert at specific times. Most noticeably, security is stepped up on the anniversary of 9/11 with a large and visible increase in armed Police on the streets of the City, in London generally, and particularly around prominent target buildings such as US-owned banks. The fear of attack from dissident Irish republican terrorists is also deemed high given problems with the peace process in Northern Ireland.

Conclusions

As a result of terrorist attack, and the risk of further bombings, a series of defensive modifications to the landscape were constructed in the City, primarily between April 1992 and February 1997. These changes were based on a radicalisation of crime prevention measures commonly employed in other Western cities in an attempt to 'design out terrorism' through the introduction of a number of armed road checkpoints, the imposition of parking restrictions, and the fortification of individual buildings. Additionally three interrelated camera networks were established and continually updated. At the end of 1996 for example, CameraWatch had 1250 private cameras, and the Police, in addition, controlled 8 permanent entry point cameras, 13 exit cameras and 47 area traffic control cameras. These numbers subsequently increased as the new extension was implemented.

The defensive strategies implemented in the City had a number of central features. *First*, there was a gradual and continual enhancement of security strategies utilised to create and maintain territorial control of selected parts of the City. *Second*, as time progressed, the 'secure zone' increased in size and status – from an amalgam of minor traffic modifications after the 1992 bomb which sought to safeguard the core financial area, to the subsequent mobilisation of the ring of steel and its extension. The processes involved in bringing a greater area under such control was aided by the increased co-ordination between businesses, and in relation to cost, legality and public opinion which had hindered the original plans to construct a cordon in 1992. *Third*, the process of making permanent landscape changes in the City was articulated not in terms of counter-terrorism benefits, but in relation to reductions on crime, environmental pollution and traffic congestion. This, although fitting in with prior planning decisions, can be seen as a deliberate attempt by the police and the Corporation to remove references to terrorism that could undermine the City's attractiveness to global finance corporations. This idea will be further advanced in Chapter 7.

Fourth, as the ring of steel grew in size its spatial effects on other areas became more apparent at a number of geographical scales related to its 'collar zone' directly outside the cordon, local effects caused by the exportation of traffic and potential terrorist risk into neighbouring boroughs, and, to other key target sites in London which, due to the heavy security in the City, were felt to be more exposed to attack. *Fifth*, the response from the agencies of security to the terrorist threat served to strengthen the links between the City police and the other agencies of security and the general business community. The approach of the City to anti-terrorism should therefore be seen as a 'joint effort' highlighting the increasing importance and co-ordination nature of risk management within modern society.

In short, the City police sought to control and regulate the space within the Square Mile, attempting to create a particular image for the City based on a balance between a flourishing business culture, and safety and security seen as vital if the area was to be competitive within the global economy. They were helped in this by the private security industry. The City embraced inclusion in the globalisation process whilst at the same time excluding themselves from the rest of central London through their territorial boundedness and fortification strategies. The situation that developed in the City can therefore be seen in terms of a condition of global connection and local disconnection – a condition, which, as a result of terrorism, continued to characterise the dislocated nature of the City's relationship with the rest of London.

In this context the ring of steel and other security measures highlight two of the most important trends in contemporary urbanism. *First*, the increased militarisation of urban space where the relationship between fear and form determines the extent of fortification measures. Initially there was a reluctance to introduce radical fortification measures to the City as it was felt this would 'stigmatise' the area and increase the fear of those 'cocooned' within the secure area. However, as the risk of further bombing increased, so did the desire of occupiers in the City to bring about alteration of the physical landscape to enhance security. *Second*, the City's defensive landscapes express the privatisation and control of space, according to the preferences of the rich and powerful, with control being increasingly asserted through physical and technological measures. This was associated with the need to maintain a predictable, efficient and orderly flow of commercial activity. The business community therefore supported enhancement of security at times of risk but wanted it downgraded when they perceived that the risk was declining.

The observations noted above should be seen as part of an overall schema that was in operation in the City due to the risk of terrorism, namely that through enhanced territorial strategies at a local level aimed at reducing the risk of terrorism, the Square Mile was increasingly able maintain its position in financial markets on a global level. The spatial imprinting of this trend was is evident through the construction of defensive landscapes, which have become a concrete part of the contemporary urban scene in the City.

Post 9/11, the ring of steel has once again been reactivated and is being used explicitly for counter-terrorism purposes. New technologies continue to play a significant role in developing both the surveillance capabilities of the City Police as well as initiatives developed for businesses within the Square Mile. The key point here is that in reality, apart from the natural upgrading of technological systems and a reassessment of contingency planning, the City of London has changed little since 9/11 given that its counter terrorist approach was already well developed and could be activated immediately.

Indeed in the early months of 2003 it was announced that the Corporation had committed itself to raise the business rate premium for City businesses in order to improve security through, for example the addition of extra cameras[50] As the Chairman of the Policy & Resources Committee noted:

> The security of the City of London has always been a priority for the Corporation of London and the City of London police and these extra funds will enable us to expand the 'ring of steel' and put more police on the streets... Most businesses we consulted were in general favour of spending more on security but were disappointed that central government could not help more...We've listened very carefully to the business community and we think we've struck the right balance: security remains a priority and we're confident that the new recommendation will still mean a significant increase in the resource we put into keeping the City as safe as we can.

The economic importance of the City and its attractiveness to terrorists was also reaffirmed in 2003 when it was reported that the threat of a major terrorist attack on London led the Chancellor of the Exchequer to consider 'radical contingency plans for the City' (Curphey, 2003). It was reported that the Chancellor was:

> Seeking the power to take control of the City in the event of a terrorist attack. The measures, which are under consultation, would help to limit damage to the UK economy if terrorists were to target the Square Mile in London with a dirty bomb or biological attack.

It would give the Chancellor the authority to freeze payment and settlement systems and run the Bank of England in the aftermath of an attack.

The article continues by noting that the powers were necessary because:

The government's fear is that terrorists might inflict even greater damage on London than on New York in the September 11 attack. If they were to release a chemical, nuclear or biological attack on the City, the whole area would have to be cordoned off for months or years. Banking staff would be unable to return to their desks, and the millions of financial transactions which normally take place each day would stop...As banks and investment companies use just a few trading platforms between them, the whole of the financial sector in the UK could collapse. This would scare foreign investors and inflict serious damage on the UK's reputation as a world player in the financial markets.

Notes

1 See for example Morris (1994).
2 This wall was more then 3km long, 2.7m thick at the base and at least 6m high.
3 These include Ludgate, Bishopsgate, Aldgate, Newgate and Aldersgate. These gates no longer exist but the names are retained in contemporary street names.
4 It is beyond the scope of this book to discuss the trading implications of the Big Bang other than to note that the financial revolution led to a realisation that traditional insurance techniques were out-of date and that re-insurance would need to be more widespread to offset the risk and avoid bankruptcy (see for example Hamilton, 1986; Welsh, 1986).
5 During this attack three bombs were launched from an improvised mortar launcher attached to a van which was parked approximately 200 metres from Downing Street. One of these bombs landed in the garden of 10 Downing Street during a meeting of the Gulf War Cabinet. The other two shells failed to explode.
6 Police Committee of the Corporation of London, 27 January 1993.
7 Police Committee of the Corporation of London, 27 May 1992.
8 This attack occurred less than a month after Sinn Fein had published *Towards a Lasting Peace*. This document outlined how Irish unity could be achieved from their perspective.
9 Also referred to as rolling random armed road blocks or 'Operation Rolling Rock'.
10 The number of officers in the CTS continued to grow in 1992 and increased again during 1993 and 1994 (Annual Report of the Commissioner of the City of London Police, 1993).
11 This was finally approved at the Policy and Resources Committee on 18 June 1992 under the following statement of reasons 'to aid traffic flow and road safety'.
12 The main committee of the Corporation of London.
13 This was related to decisions made at the Court of Common Council (25 June [also 16 July]). The Court of Common Council on 10 September also decided that further traffic management measures related to loading, parking and waiting restrictions, which would complement and offset any adverse reaction to the implementation of previous restrictions, should be implemented.
14 See Kelly (1994b).
15 This device killed one person, injured 94 and caused considerable damage initially put at £1 billion (subsequently reduced to around £550-600 million).
16 The newspaper of the Republican Movement in Ireland.
17 *An Problacht/Republican News* (29 April).
18 See Sivell (1993).

19 Cited in *The Independent*, 5 May 1993.
20 Cited in Ford (1993a, see also 1993b).
21 Subsequently, legal changes in the Criminal Justice and Public Order Act provide much needed extensions to PACE which allows officers to search where he/she thinks fit with no restriction placed on suspicion. This Act received Royal Assent in November 1994/January 1995 (see for example Hill, 1995).
22 Cited in Burns (1993).
23 The term ring of steel has entered the British vocabulary in the 1990s due to a number of high-profile terrorist attacks committed by the Provisional IRA in England. The popular definition of this term has now been blurred by the media and is used to refer to any area that undertakes a high profile police presence in order to deter a terrorist event.
24 Cited by Garvey (1993) in the *Evening Standard*, 29 April 1993.
25 See minutes of Policy and Resources Committee (11 November 1993).
26 In addition, after Bishopsgate the methods by which the vehicle checks were made in the collar zone were changed as a result of the lessons learnt by an officer from the City force being seconded to the RUC in Belfast, with the efficiency of vehicle checks improved to allow more to be carried out using the same manpower.
27 25 September 1996.
28 The extension to the cordon as will be shown, was actually planned in 1995 and came into effect in early 1997 and later in 2000 in order to protect this important and valuable new area of the City.
29 At the Police Committee on 26 July, and agreed at the Court of Common Council on 9 September, it was also proposed that colour units replace the existing monochrome cameras. These would require the installation of additional monitors at Wood Street Police Station. The total cost was estimated at £473,000. This equipment was installed by the police by the end of February 1994 (Police Committee, 26 January 1994). At this time it was decided to install four additional cameras to cover the Upper Thames Street tunnels at an additional cost of £120,000 (see also Notton, 1997).
30 Cited in the 1993 Annual Report of the Commissioner of the City of London Police (1994).
31 See 'CameraWatch' (Commissioner of the City of London Police, 1993).
32 Cited in the *Electronic Telegraph*, 5 June 1997.
33 There are other precautions that can be taken to reduce the effects of a bomb blast such as application of window film, installation of blast curtains and the construction of shelter areas. These measures however lie outside the scope of this project as they are related to the interior of the building and not the landscape of the City.
34 Minutes of the Police Committee, 2 February 1995.
35 It is further reported by City Police that after the Docklands bomb in February 1996, and a reactivation of the full ring of steel, crime was reduced substantially (Police Committee, 8 January 1997).
36 These were installed (at a revised cost of £510,000), 13 further cameras were added to monitor the cars exiting the City at a cost of £323,000. Before this date security cameras only focused on cars entering the cordon. Exit cameras were particularly important as police could now monitor traffic into the City and if needed track suspect vehicles exiting the cordon and new higher resolution cameras were added to the entry points at a cost of £230,000 (Police Committee, 29 March 1995).
37 See the City Engineers Report, 26 September 1996 and the *Electronic Telegraph*, 22 January 1997.
38 See Binney (1996).
39 As such on May Day 2000 when similar demonstrations were planned a highly organised joint operation was put into play.

40 The Real IRA was born out of a split within the ranks of the Provisional IRA in October 1997.
41 *City Security*, 9/2001, p.4.
42 Minutes of the Planning and Transportation Committee, 29 February 2000, and 11 April 2000.
43 Minutes of Policy & Resources Committee, 29 July 1999.
44 Minutes of the Planning and Transportation Committee, 4 July 2000.
45 Minutes of the Planning and Transportation Committee, 17 October 2000, and 20 March 2001.
46 Cited in *City Security*, 9/2001, p.4.
47 City of London Police press release, 14 November 2001.
48 Although concern was expressed about 'unscrupulous individuals playing upon peoples' fears by attempting to sell protective clothing and gas masks to City businesses.
49 Cited in *This is London (Evening Standard* online), 8 October 2001.
50 Corporation of London news release – 'Committee recommends new business rate premium for City policing and a bigger "ring of steel"', 18 February 2003.

Chapter 6

Distributing the Financial Risk of Terrorism

Introduction

The previous chapter introduced the strategies employed by the agents of security as a response to the terrorist threat and the resultant alterations to the physical form of the City's urban landscape. Increased fortification provided material evidence that the City did all they could to restrict the damage the Provisional IRA could inflict on the UK economy through attacking the reputation and physical infrastructure of the Square Mile.

A less obvious, but equally important, manifestation of the reaction to terrorism in the City, especially during the 1990s was the establishment of a terrorism insurance scheme to cover damage caused by terrorist bombs. In this regard risk was viewed not as a physical threat to be dealt with, but in terms of statistical tables, relating to the probability of premium losses and premium gains in certain locations in relation to the threat of terrorism. As such insurance coverage operated as a risk spreading mechanism and served to create distinct distributive geographies based on the principle of financial exclusion or 'redlining'.

This chapter will highlight how the insurance strategies employed in the decade after the 1992 bomb sought to manage and control terrorist risk in two main ways. First, through attempts to distribute the financial risk to terrorism throughout the national and global market place in order to remove liability from certain locales (especially the City); and second, by drawing boundaries around high risk 'hot spots' and encouraging, or insisting upon, risk management measures and contingency planning as a condition of granting insurance coverage. This chapter will also highlight how the methods of financial risk spreading employed by the insurance industry were of limited success, succeeding only in concentrating the financial risk of further bomb attacks in the Square Mile. This will highlight implications for the geography on the ground by analysing how the City, on one hand, attempted to redistribute its localised risk in the national and global economies through insurance mechanisms, whilst on the other hand it increasingly fortified its urban landscape.

When the Provisional IRA stepped up its campaign of economic disruption in England in the early 1990s questions regarding the insurability of terrorist risk were being asked at the highest levels of commerce and Government, especially given the importance of the City, in the global economy. These are questions that have once again been asked post 9/11 as the insurance industry has responded to the new climate of terrorism.

The remainder of this chapter is divided into three main parts. First, it will analyse the methods and success of the risk spreading mechanisms employed by the insurance industry in relation to terrorist risk in the 1990s. The second part will link the provision of insurance to changes in defensive landscape features, indicating the subtle relationships between these two methods of 'risk management'. The third part will highlight the impact of 9/11 on terrorism insurance mechanisms both within the UK and global markets.

In relation to the provision of terrorism insurance the period since 1990 can be summarised in six stages:

- Emergent risk (1990 - April 1992);
- Reflecting on local terrorist risk (April 1992 - November 1993);
- Transferring the risk nationally (December 1992 - April 1993);
- Concentrating risk in the City (April 1993 - August 1994);
- Fluctuating risk and alternative risk sharing mechanisms (August 1994 - September 2001);
- Globalisation of risk (September 2001 - present).

These stages will now be examined in turn, and related to the spatial processes operating within, and in relation to, the City.

Emergent Risk - Reduced Market Capacity But Increased Risk

During the early 1990s there was growing concern in the insurance market that the solvency of insurance companies could be threatened as a result of both natural disasters and old financial liabilities which were increasingly coming to the fore. In the City, the consequences of the latency of financial risk can be exemplified through insurance losses experienced by Lloyds of London in the late 1980s. Between 1988 and 1990 a string of natural and technological disasters such as the fire on the Piper Alpha oil platform, the oil spillage from the Exxon Valdez and a series of earthquakes and hurricanes in the United States meant a number of large insurance claims were sought against the company. Two features of the market compounded this situation at this time. First, insurance premiums were dropping as a result of increased corporate competition. Second, many 'old' risks were being claimed against, such as asbestos-related illnesses, genetic engineering and many different types of pollution. Insurance companies therefore began to closely examine the risks they were prepared to underwrite.

The periodic bombing of London at this time by the Provisional IRA meant the insurance industry was unable to predict the likelihood and potential costs of such attacks, meaning that consideration was given by insurance companies to withdrawing from the terrorism insurance market.

As early as October 1991, the Association of British Insurers (ABI) were in discussions with the Government to review the existing cover against terrorism, as during the first Gulf War it was far from clear who would be financially responsible for associated acts of terrorism in the UK. This argument was important, as at this time terrorism was covered within standard insurance policies without additional costs to the premium holder. This could have had widespread geographical implications in some areas, such as the City (and other areas seen as economic targets by the Provisional IRA), if insurance against terrorism was unavailable as a result of rising threat levels. It was therefore realised by the insurance industry – themselves based predominantly in the City – that continued risk spreading through the international reinsurance market was a necessity to protect their financial liability. This threat was fully realised in April 1992 at St. Mary Axe. This bomb brought about a chain of events, which led to attempts to re-distribute the financial risk of terrorism at a variety of spatial scales away from the City in order to protect its insurance liability as well as the trading position of the Square Mile.

Reflecting on Local Terrorist Risk

In Northern Ireland, the Government had long accepted responsibility for damage caused by terrorism given the withdrawal of traditional insurance mechanisms in the late 1960s (Greer and Mitchell, 1982). The development of the Northern Ireland 'compensation scheme' can be traced back to the 1950s and 1960s, during which insurers dealt with many of the same issues that confronted UK insurers in the early 1990s. These included:

- The insurance industry effectively withdrawing from the market;
- Compensation being paid by the state and hence the financial risk being redistributed nationwide with the premiums of the many paying for the losses of the few (essential through a slight rise in tax);
- The development of specific anti-terrorist measures becoming necessary as terrorism insurance became increasingly politicised due to Government involvement, and areas of high economic importance became increasingly attractive targets.

After the St. Mary Axe bomb in 1992 the insurance industry became very proactive and began an immediate review of their cover of terrorism losses. Similar arguments to those used in Northern Ireland in the 1970s were put forward by the insurance industry to try and get the Government involved and remove their own liability, believing that the Northern Ireland scheme set a precedent of Governmental participation in providing insurance coverage against terrorism.

An Increased Threat Level?

After the St. Mary Axe bomb, figures released by the insurance industry, and advanced through the media, noted that the cost of damage (initially estimated at around £800 million) was likely to be more than the total cost of damage in Northern Ireland over the previous 22 years (around £600 million). This publicity, the police believed, increased the threat level to the City. This bomb, according to the insurance industry, demonstrated the potential insurance cost of a major strike against the City, bringing home to the Provisional IRA the operational effectiveness of planting a bomb in Britain's financial centre, given the disruption it caused to the British economy.

Subsequently it emerged that the initial figures released by the insurance industry were a large over-estimation. Some saw this as a deliberate attempt by the insurance industry to get the Government involved by overemphasising the magnitude of the problem, as they saw the issue of terrorism insurance as the responsibility of the state, and not commerce. As one leading insurer indicated "it may have suited the insurance industry not to analyse the expectations of loss too closely as they were keen to get the Government involved". The net result was that both the British insurance industry and their reinsurers began to publicly express concern about their future liabilities to such risk, and hence their ability to underwrite terrorism insurance in the UK. The ABI voiced its concerns to the Home Secretary in May 1992, to the President of the Board of Trade in June and to the Department of Trade and Industry (DTI) in July:

> It is a social risk but it is also a political risk. However there was an increasing corpus of feeling that the Government should have taken more responsibility for this.
>
> (Senior City Insurance Consultant)

The insurers concerns were for two main reasons. First, the large number of catastrophic incidents that had occurred in the years preceding the bombing, which had pushed many insurers and reinsurers near to the limits of insolvency. Second, the nature of terrorist attack defied most of the normal 'laws of insurance', as the insurance industry could not quantify the potential financial exposure of a terrorist bombing when they could not predict where it is going to be located, its explosive force, or how business disruption would effect financial markets. It was also reported that the reinsurers realised that the direct insurers did not have accurate financial risk profiles in place for areas, such as the City, and hence adequate information about potential liability could not be accurately calculated. As such given these restrictions the insurance industry felt economically vulnerable:

> If the UK insurance industry had continued to underwrite terrorism risks, and the bombing campaign had continued at a high level, the industry could potentially have gone bankrupt.
>
> (Association of British Insurers Director)

There was also obvious annoyance from the insurers that the Government were, at this time, unwilling to support the market either by extending the Northern Ireland scheme to the British mainland, or by financially supporting the existing market. Government reluctance was for two main reasons. First, it reflected a particular concern over cost, given that the bill for St. Mary Axe was estimated as equivalent to the sum of twenty years of terrorism-related compensation paid out in Northern Ireland. Second there was a feeling that if the Government subsidised insurance coverage the terrorists could increasingly claim that attacks against economic targets were politically motivated. Therefore the Government, in contrast to the insurance industry, saw the issue of terrorism insurance as an issue for the insurance market.

The argument the insurance industry continued to highlight was that if the Government paid for the damage caused by terrorism it would come out of public funding and the whole community could have contributed a small amount in taxes. Not unsurprisingly, the view of many in the City was that the Square Mile should not have been singled out for higher premiums, but instead, the cost of insuring society against terrorism would be equally distributed in line with other risks such as flooding. This would have redistributed the risk, in financial terms, away from the City institutions and their international reinsurers. As a senior City insurer noted: "It shouldn't have just been lumbered on those people who pay an insurance premium in a particular area. It should be more equitably spread over society."

Alternatively, some proposed that the Government could have imposed a compulsory levy against commercial companies (see for example Bagnell, 1993). The Government at the time was against charging a compulsory levy to commercial organisations, as companies in low risk areas would see it as unfair if they were subsidising the more attractive target areas. This was seen as a politically unacceptable solution:

> Another way of dealing with the risk would have been through a compulsory levy for all insurance premiums. In this situation all those who buy insurance are contributing to a common pool, which would pay for the losses of the few – the general principle of insurance. This was undoubtedly considered but was rejected as it would have created a high profile situation for the Government.
>
> (City Insurance Broker)

A Northern Ireland Model for Britain?

Several months after the St. Mary Axe bomb the insurance industry was still concerned about its liability and was still putting pressure on the Government to introduce a similar scheme to that in Northern Ireland. The Government were not keen to extend the Northern Ireland model for both financial and political reasons:

> It felt like Northern Ireland, in that the Government should take the risk but the Government were adverse to taking the risk, not just because they didn't want to take it financially, but at that stage they were absolutely petrified of being seen to have bowed to terrorism and taking it in the public purse.
>
> (Senior Risk Manager)

The Corporation of London, too, saw the Northern Ireland model as the obvious solution to go for. As Michael Cassidy, Chairman of the Corporation's Policy and Resources committee, stated:

> Because we were in new territory, and it was simply my job to make sure something happened, the most obvious model to go for was the Northern Ireland scheme.

Not only would a Northern Ireland scheme have been the easiest to establish, but it would also have been of most benefit to the City, as it would completely remove any financial liability from insurance companies within the Square Mile – as well as the easy option given the panic that was growing with the immediate need to come up with a suitable scheme. Cassidy indicated that the Government were "very resistant to any idea of extending the Northern Ireland scheme" as this would be viewed as politically unacceptable at the time. A Corporation of London insurance officer also indicated that he felt the Government were not keen to extend the Northern Ireland scheme at this time due to the fragile majority they had in the House of Commons. As he noted "Insurers were hoping for that...consumers were hoping for that...but the Government were not keen as they didn't want to increase their tax burden at a time when the Government were in some difficulty".

Circumstances came to a head in October and November 1992 when one of the major European reinsurers wrote to all their ceding companies[1] indicating that they would be excluding terrorism from their standard policies from January 1 1993. Due to the potential financial liability from terrorism it was inconceivable that any particular insurance company would be willing to bear the whole risk themselves. In normal circumstances a reinsurance arrangement would have been entered into where a company does not accept the whole risk, but parts of it with other insurers or reinsurers through the automatic transfer of risk via a standard 'contract' which covers a certain percentage of every risk an insurer undertakes on a certain risk. The biggest reinsurers in the world at this time were European based, predominately located in Germany and Switzerland, and without their support it was impossible to obtain cover at economically viable terms for any major terrorist-related risk. This meant the underlying insurers who relied on this reinsurance cover would not offer terms of cover without this back-up facility. Therefore, when the large reinsurers removed themselves from the market, the British reinsurers subsequently followed suit, and in turn direct insurers said they would operate a terrorism exclusion from their standard policies.

The ABI, at this time, didn't really have to do anything to resolve this situation as the onus was on the Government:

> In reality they [the ABI] didn't need to do anything because they were walking away from it and therefore the insurance market didn't have to necessarily have to try and find another solution.
>
> (Senior City Insurance Consultant)

It was, at this time, that bodies responsible for the organisation of risk management and insurance in Britain such as the Association of Insurance and Risk

Managers in Industry and Commerce (AIRMIC), the Confederation of British Industry (CBI) and the British Insurance and Investment Brokers Association (BIIBA) began to get involved with the ongoing debate:

> In October 1992 before the exclusion was announced I saw some correspondence from Munich Re, one of the two biggest reinsurers in the world, indicating that they were going to operate a terrorism exclusion from the year end. Nothing had been said at this point. I took this to AIRMIC and we leaked it to the *Financial Times*.[2]
>
> (Senior AIRMIC Consultant)

The involvement of these insurance and risk management groupings typifies how the assessment of risk is now determined in modern society, as financial necessity has meant an ever expanding range of institutions now seek to judge risk for their own benefits. In particular, the involvement of these insurance and risk management groupings could be seen as an attempt by these bodies to establish a 'voice' for themselves in order to strengthen their position. In addition, the ongoing debate about the removal of insurance cover could also be seen as an event which stimulated the Corporation of London to get involved in this matter.

The removal of reinsurance cover also caused much concern amongst businesses in locations seen as vulnerable to attack, especially given the expiry date of many insurance policies on 31 December 1992. The situation was particularly worrying for the Corporation of London who were due to renew their own insurance policies even earlier on Christmas Day. A Corporation's insurance officer indicated that they had heard rumours from as early as May 1992 that the reinsurers were going to pull out of the market. This caused concern in the City. As Michael Cassidy noted:

> We [the Corporation of London] felt that this was a national Government responsibility and that to leave people unable to obtain cover was going to be completely unacceptable to the City so we tried to mobilise the Government into doing something.

Both the insurers associations and the Corporation of London were keen to get the Government involved as they saw the risk of terrorism as an issue which affected the whole country, and believed the distribution of financial risk should reflect this instead of being concentrated in certain areas such as the City:

> There was a lot of lobbying from the Corporation of London for two reasons. Firstly because they found themselves with the risk that their own property portfolio, which is enormous, was not covered. Secondly the City as a whole would not be covered. Both of these thing drove the Corporation to lobby very hard indeed.
>
> (Senior City Insurance Consultant)

In short the City's localised response was an attempt to reduce its vulnerability against the financial implications of a bomb which would not only be detrimental to its long term prospects as a global trading centre but would feed back to affect the entire UK economy.

The ABI Standard Exclusion

On 12 November 1992 the ABI issued a press statement indicating that it had advised its members (the majority of the UK insurers) to exclude terrorism from its industrial and commercial policies in line with most other European countries. The statement went on to blame the reinsurers for this scenario in that 'leading world-wide reinsurers have forced this exclusion on the UK market in the light of considerable losses earlier this year in major bombing incidents and the continuation of terrorist bombings in the UK'.[3]

The international reinsurers still insisted that Provisional IRA terrorism was a political issue that the British Government must address, just as it did in Northern Ireland in the 1970s. Other commentators pointed out the coincidence of the press release, occurring just after a bomb was found, and defused, at Canary Wharf in the London Docklands, which would have caused massive devastation if detonated. Gloyn (1993), for example, noted that this near miss did nothing to weaken the resolve of insurers to pull out of the market for terrorism.

A period of stalemate followed where the ABI appeared to want to wash their hands of the issue whilst the Government were equally determined that they were not going to be bullied into submission and portrayed the issues as a commercial matter. The timing of the ABI announcement could not really have been worse as there was much anxiety surrounding the fear of Provisional IRA attempts to bomb London, and particularly the City over the Christmas period:

> In the insurance market there was a great worry about terror insurance with Christmas coming up and the understanding that the IRA are likely to try and give a present [a bomb] to the City before Christmas. The Corporation were very nervous.
>
> (Corporation of London Insurance Officer)

Others reaffirmed that the insurance ruling was, in their opinion, a boost for the Provisional IRA as it made the campaign to attack economic targets more attractive, as well as acting as a disincentive for firms wishing to relocate to Britain, and in particular London:

> My view was that if you go back to basic terrorist philosophy the IRA wished to compromise the economic well-being of the country, so, it could be argued that once insurance cover was reduced, or withdrawn, the terrorists have won – we have actually given them an economic victory.
>
> (Senior City of London Police officer)

Reducing the Competitive Position of the Square Mile

At this time the Lord Mayor of London wrote to the Prime Minister expressing concern over the future economic success of the City if the Government refused to offer cover against terrorist attack. Indeed, an internal memorandum circulated by the Corporation of London on 24 November 1992 indicated that the City's global reputation could be jeopardised by the bomb, noting that 'if the situation were unresolved, taking high grade property in the City would be a less attractive [i.e. risky] option'.

The situation could have had disastrous effects on London's economic competitiveness through impacts on the general economy, by reducing the attractiveness of locating in the City, and by affecting the investment potential of City institutions. *First*, it was felt that lack of adequate terrorism insurance could harm the already fragile economy which was just coming out of recession, by a combination of the threat of terrorism and the unavailability of insurance which would weaken the commercial position of the City, and the UK as a whole. In particular there was a fear that the removal of terrorism insurance could discourage businesses from locating in the Square Mile and led firms to relocate in other European cities, which were in direct economic competition with the City. There was also concern that the situation would jeopardise the efforts at this time to encourage the Central European Bank to locate in the City.

Second, there were concerns that the property investment industry could be affected as property developers would not be able to secure funding on the strengths of the property they owned because it would not be possible to fully insure their stock. The lack of financial security could also have led to a future lack of investment property development. For example, chartered surveyors in the City pointed out at this time that the inability to obtain terrorism insurance was delaying the completion of property deals. Similarly, there were fears that tenants who would have to bear the cost of terrorism insurance through their leases would avoid high premiums and move elsewhere.

The lack of terrorism cover therefore increased the vulnerability of high profile City buildings to attack, with implications for the future functioning of the City's economy. For example, a letter to the Prime Minister from the City Property Association[4] underlined that, 'the City is clearly a prime target, not only because of the value of its buildings, but also because of its importance to "UK LTD"'.

The letter continued by indicating that the Government should step in to help avoid 'areas of the City of London remaining literally as "bomb sites"' and to stop the City 'becoming a "no-go" area in terms of occupancy and investment, to the benefit of our overseas competitors, particularly in financial services':

> It was important to the City that business continued and didn't go to Frankfurt or anywhere else. The fact that the London insurance market couldn't buy commercial reinsurance for itself meant that they were unable to offer primary insurance cover for bomb damage. This had big implications for business.
>
> (Representative of Pool Reinsurance)

Transferring the Risk Nationally

After the ABI announcement, the Corporation of London took the lead in trying to persuade the Government to get involved with covering terrorism insurance and help spread the financial risk nationally, avoiding a situation where the City's niche in the global market place was adversely questioned. A letter sent by the Lord Mayor of London to the Prime Minister on 2 December 1992 noted that:

> In our view, the withdrawal of insurance facilities will discourage new businesses from locating in the City and could well be the catalyst for foreign institutions to relocate elsewhere.

Furthermore this letter highlighted that if the position of one of the inter-related financial markets was jeopardised this would feedback into the overall performance of the City:

> London is successful because its various markets and professional services feed off each other and this fragile relationship could be harmed if there were to be a loss of confidence.

The letter continued by indicating the link between the lack of available insurance cover and the development of an increasingly fortified urban landscape:

> We are already aware that some property owners are proposing to initiate their own security arrangements in order to protect themselves against an open ended risk and the appearance of security measures such as fences, etc alongside our streets will further undermine confidence.

This is indicative of the relationship between financial security provided by insurance and physical risk management measures utilised to reduce exposure to particular risks. It highlights the different ways in which insurers viewed terrorist risk at this time. On one hand they reduced their own risk exposure in financial terms as they signalled their intention to withdraw from the market. On the other hand, the insurers had served to enhance the physical risk level in the City as the non-availability of insurance cover made it an increasingly attractive target to terrorists, and hence, encouraged businesses to install additional risk management measures in an attempt to reduce their potential liability costs.

Pressure Continues to get the Government Involved

In the lead up to Christmas 1992, the Corporation of London continued its efforts to get the Government involved. In a press release on 4 December 1992, Michael Cassidy noted that the City was anxious about the terrorist insurance issue and that 'the Government must resolve the insurance position to ensure that a framework is provided in which business can proceed with confidence'.

By mid-December, the media had at last begun to realise fully the implications of the removal of terrorism insurance cover given the stalemate between the insurers and the Government. A number of small bombs in London in early December had no doubt contributed highly to this state of affairs. For example, a City underwriter indicated that the media were now beginning to show an interest in the story, which served to publicise the Corporation's case:

> One or two newspapers at this time started to take up the theme sensing that there was something quite big here. This no doubt helped the pressure wave that was building on the Government to sort the situation out.

The Corporation in their desperation to persuade the Government to act as 'insurer of last resort' even enlisted the help of the Prince of Wales who indicated that he would raise the matter personally with the Prime Minister. This followed comments by Michael Cassidy, that some of London's prestige landmarks could not be rebuilt if destroyed by bombs (see Olins, 1992 and Carolan, 1992). Furthermore, a confidential letter from the Lord Mayor to the Prime Minister indicated that until the insurance situation was resolved 'German and Japanese investors are no longer prepared to purchase property in the City'.[5]

However, the ABI continued to insist that the insurance industry could and would not act without Government support. What, however, was clear was that the negotiations that took place between industry representatives, the Corporation and the Government were vital to creating pressure which forced the Government to look at its stance. Indeed, meetings the, Corporation had with the Government in December 1992 were, he believed, 'instrumental' in getting the government involved:

> It is true to say that the Corporation was a major factor in getting the Government involved - using the argument that with the country's economy the City is a big part, and that in this regard, they [the Government] need to provide a safe place for domestic and overseas financial institutions to exist within the City boundaries.
>
> (Corporation of London Insurance Officer)

The Government as Insurer of Last Resort

On 21 December 1992 the Government amended its position indicating that it was willing to act as the ultimate reinsurer, the so-called 'insurer of last resort', behind a Pool of Insurers who had agreed to set up a mutual company, Pool Reinsurance (Pool Re), to provide cover in the traditional way against terrorism. Under this scheme the insurance industry were effectively passing on all the additional premiums in return for the transfer of the terrorist risk to the Government who were reinsuring the scheme. As the Department of Trade and Industry (DTI) press notice stated, 'in recent months it has become virtually impossible for insurers operating in the UK to obtain the reinsurance protection they need so they can continue to offer terrorism cover in Britain for new policies and renewals from the end of the year'. The statement continued: 'In the light of this decision the Government expects that owners and tenants of industrial and commercial property should continue to be able to obtain cover against terrorist attacks. The price of such cover will, however, be adjusted to reflect the changes in the risk in the usual way.'

The scheme was a compromise between a Northern Ireland model, which was dismissed as politically unacceptable, and a commercial solution, which the insurance industry had rejected. This scheme represented a series of promises by the Government, which in due course would be ratified by parliament. The formation of

Pool Re averted a potential crisis in the property industry as tenants and owners were now able to purchase insurance cover against terrorism, albeit at great cost. The way that the scheme was set up ensured that the Government would not be held responsible for compensation for businesses who are not insured for terrorist risk, and that the premium rate for such terrorism insurance policies would be periodically reviewed in line with developments in Northern Ireland. The Government agreed to meet 90% of further claims not covered by premiums gathered by Pool Re. The insurance companies would collectively cover the remaining 10%. The Government at this time also insisted that their involvement would be short term and would not be detrimental to the taxpayer.

Government involvement effectively aimed to spread the financial risk of further terrorist attack throughout the national economy, and away from the City markets. This reduced risk as the Government, given its powers, could absorb the potentially huge cost of terrorist attack. In addition, Government involvement effectively stopped other insurance or reinsurance companies, both home and abroad, who might have considered underwriting insurance from doing so, as there was no incentive for them to compete with the British Government given the enormity of potential losses.

Insurance Zoning

The Pool Re scheme had a number of unforeseen geographical impacts, which meant, that the Government's aim of national risk spreading was defeated. In particular the risk became spatially concentrating in the selected urban areas which were seen as key terrorist targets. Under Pool Re, areas of the country were designated as either high risk (Zone 1) or low risk (Zone 2). This designation was for two reasons. First, the speed at which the scheme was set up required a simple, yet workable, premium rating structure. Second, by setting the scheme up with only two zones meant that all those in Zone 1 would be charged the highest premium and hence, if there was a large take-up of policies, the pool of premiums could be maximised. This logic relied on all those businesses in Zone 1 regarding themselves as 'at risk' from terrorism.

There were nine designated high-risk centres – Birmingham, Glasgow, Leeds, Manchester, London (not just the City and West End, but all London postcodes), Bristol, Liverpool, Edinburgh and Cardiff. High-risk areas had substantially higher premiums, which reflected their greater perceived risk exposure. In addition, if property was located next to a high-risk property, such as a Government building or a high profile construction project, an additional 'target risk' premium of 50% was charged. In short, Zone 1 rates were approximately 3-5 times those of Zone 2 without the target risk classification. Zone 1, in the opinion of some, covered areas of the country which were felt to be at little risk from terrorism. As a City insurance company director indicated, this classification was particularly hard on non-London areas in Zone 1 as 'there may have been a number of areas seen as high risk but we all know that the City and perhaps Oxford Street or Knightsbridge are the realistic targets'.

The perception at this time was that London was the main target of the Provisional IRA, and that the other 'high-risk' rated areas were put in Zone 1 to increase the pool of money the scheme collected. City of London premiums were especially high due to the large number of buildings within the Square Mile considered 'target risks'. However, Pool Re was still advantageous for the City, as their risk was in effect subsidised, at least in part, by businesses in other locations. Such territorial rating in this case could be seen as an attempt to let 'good' risks subsidise 'bad' risks in line with the general principle of insurance.

However, the Corporation still continued to push the DTI for reduced premiums. Michael Cassidy noted that he was annoyed that the City was being singled out for higher premiums:

> The Corporation of London is not the only area of the country at risk from the terrorists, the whole of the country is. We felt that we should not receive higher premiums as a result.

The Corporation were still disappointed that insurance costs would, in their opinion, unduly affect City properties. They saw the City as a national economic asset, which was being plagued by a national fight against terrorist attack and that they would like to see 'a national scheme to deal with a national battle' (Cassidy). The Corporation, in this sense, was trying to persuade the Government to spread the risk even further by setting premium levels the same for all areas of the country, which would have led to reduced premiums in the Square Mile. However, realistically the City at this time was considered one of only a small number of areas the Provisional IRA would actually target. Most of the country was not at genuine risk from significant explosions and would therefore be unwilling to give patronage to the scheme.

However, the Government's decision to finally become insurer of last resort was reluctantly welcomed by the City. The Corporation of London was pleased, yet cautious, about the announced scheme. As Cassidy stated: 'I think it's very good that the Government have seen sense on this and are able to respond in the way they have but there are many details to be resolved' (*Newsroom South East*, BBC1, 21 December 1992). As this news report highlighted:

> The City of London's breathing a sigh of relief tonight, with the Government's decision to underwrite insurance cover against terrorist attacks. Insurance companies had threatened to withdraw cover next week...The Government's initial reluctance led to warnings of economic disaster, not only for the City but the UK as a whole.

Bernard Harty, the Chamberlain of the Corporation of London, indicated that the announcement 'must be a mixture of concern that premiums will increase so much – in our case it could be somewhere between 90 and 100 per cent which is very significant. But, on the other hand, up until today we had no prospect of cover of any kind'.[6] Insurance, in this sense, was seen as a very important safety net against the risk of further terrorist attacks in the City.

At this time there was even talk that the Corporation had agreed a special discount from Pool Re for insuring all its property holdings in the City (some £5-6 billion worth), but specific details have not been revealed. What the Corporation continued to do, at this time, was to put pressure on the Government to negotiate a better deal for occupiers in the Square Mile. As Cassidy noted:

> All property owners need to be alerted to the fact that by pestering their insurance companies that their buildings are not at risk, and by increasing security, they can reduce their premiums.[7]

Concentrating Risk in the City

A major criticism of Pool Re after it was initiated was that the risk of financial loss from terrorism attack was not, despite its intentions, dispersed throughout the national and global economy in the way other catastrophic risks were, and hence, risk became concentrated in the Square Mile. Potentially this could have had devastating effects on the economic competitiveness of the City in terms of loss of business confidence, and in particular a strike against an uninsured building.

This concentration of insured risk occurred for a number of reasons. First, most of the members of Pool Re were British, therefore in the event of a major terrorist act the cost would be almost entirely borne by the British economy, or, in the event of the pool running out of capital, the British Government. This was in contrast to other catastrophic loss events which are dispersed throughout the global economy by reinsurance arrangements, reducing the effect on any one nation. Second, the risk spreading was cut down even more by requiring the member companies of Pool Re to act as reinsurers to the scheme in terms of a ten per cent levy they would be required to pay on any outstanding claims if the pool became exhausted. Thirdly, the risk to the British economy was accentuated as the major commercial insurers were significantly represented in Pool Re, and especially in certain geographical areas. The risk of loss is therefore not only concentrated in the British economy but in selected companies. The degree of risk spreading was further decreased by the fact that the majority of the risk was located in the City, which is insured by the same groups that fund Pool Re.[8] As Bice (1994, p.456) noted:

> If a bomb explodes in that area, the loss will not be spread throughout the country, but will be borne primarily by the insurers in the City of London. Hence risk distribution is defeated.

The insurance idea that the losses of the few should be shared by the premiums of the many broke down in this case, as owners of property outside of London were reluctant to buy terrorism insurance. As such the high rates nationwide meant that many decided not to buy cover, which meant the pool of money was very much lower than anticipated. This in part was to do with the 'tactics' of the DTI, who, in order to spread the risk, insisted that companies insure their entire property portfolio, and not just those buildings in the City or other high-risk locations. This 'all or nothing' rule was to ensure that the Government had a balanced portfolio of risk that was not just in high-risk zones. In reality, companies with exposures in many locations often chose to self-insure or take a chance by not insuring at all. Thus the size of the pool was diminished and concentrated with predominantly high-risk exposures in certain geographical areas. For example a survey carried out by AIRMIC a few months earlier indicated that 26% of their members were not planning on buying additional terrorism insurance policies as premiums were just too high (Lloyd's List, 26 April 1993).

As such the resulting situation was criticised by the Corporation of London and other economists as it did not spread the risk properly and left the City with the burden of liability where risk was disproportionately concentrated.

This links to Beck's notion of 'loser regions', created as a result of inequitable distribution of insurance provision and levels of risk. In such a situation new landscapes often develop through the actions of local institutions and organisations, which attempt to form a risk adverse environment by enhancing risk management in order to reduce risk. The defensive landscapes constructed in the City exemplify this philosophy. It can therefore be argued that this higher economic exposure became one of the catalysts for enhanced security measures to be constructed in City, especially after the 1993 Bishopsgate bomb, as the insurers, in their quest to provide financial security to businesses operating in the City, attempted to improve the physical security in areas that were most financially liable in the event of a terrorist attack.

The Insurance Response to Bishopsgate

The Bishopsgate bomb inflicted heavy damage and disruption on City businesses and put the Northern Ireland question on the top of the Government's public policy agenda. Initial reports suggested the damage from the bomb would cost around £1 billion. The decision of reinsurers to withdraw from the market six months earlier was vindicated by this attack.

Immediately after the Bishopsgate blast there were forecasts that insurance premiums for terrorism would rise dramatically as it was felt that future premiums should reflect the perceived risks on the ground. It was thought that the Government would have to pay most of the bill, as it was believed that the terrorism insurance money collected by Pool Re only stood at around £300-400 million. This estimate meant that it was feared that the taxpayer might well be asked to pay between £25-50 per household to make up the shortfall. This scenario was blamed on the reluctance of provincial businesses to take out the expensive terrorism cover.

Soon after the 1993 bomb, the initial estimates of £1 billion were being viewed as 'wildly inaccurate' by the ABI and 'a knee jerk reaction' by Lloyds (Smithers, 1993, p.3). Subsequently estimates made several days after the bomb put the damage significantly lower at between £500-800 million. As with the previous bomb at St. Mary Axe the insurance industry could be seen to be creating an impression, through exaggerated claims of damage, that they could not cope with the terrorist threat and that the Government should continue to take the burden of risk and perhaps remove any remaining liability from them. There were also renewed calls at this time for the Northern Ireland model of terrorism insurance to be adopted instead of Pool Re.

Such inaccurate forecasts, advanced through the media, were seen by some to unwittingly provided a morale boost the Provisional IRA, who could use the impact on the British economy as useful propaganda material:

> The bloody media got it wrong again. You would have thought that after the last bomb they would realise the cost is never as much as the wild speculative guesses being banded around. £1 billion is just a nice figure for them to latch on to and show the extent of the terrorist damage.
>
> (City Insurance Broker)

It can be argued that the media, through inaccurate reporting, served to construct a enhanced risk profile in the City, whilst giving the Provisional IRA a morale boost in relation to the success of the bombing.

Subsequently in May 1993 the Government provisionally published the Reinsurance (Acts of Terrorism) Bill which gave legal authority to the promises made by the DTI in December 1992. There was concern that with the formal recognition of this Bill the Government would try to recover much of the cost of the Bishopsgate explosion through raising premiums to protect themselves financially in the event of further strikes.

Subsequently, Pool Re came under immense pressure from the Government to raise premiums under a 'rate review' that was initiated following Bishopsgate. In general insurance practice, one way of reducing the risk a company bears, is by decreasing the financial consequences of that risk by charging an excessive 'safety loading' on premiums for that risk – the amount collected by the insurer which is in excess of an economically efficient premium. It is, in effect, their safety net. The safety loading increases the premium pool which can be used to pay claims. In a free and competitive market this is seldom used, as policyholders would change insurer to obtain cheaper rates. However, with regard to terrorism insurance in the UK, the absence, at this time, of any credible alternative scheme meant that the Government was considering a high safety loading. Premium increases of 300% were feared in the City. Many felt that this would further undermine the scheme and would discourage people from taking out terrorism insurance. In short it was felt that more people might 'take a chance on terrorism' or perhaps relocate their businesses away from high risk zones. The ABI commented that already some twenty to thirty per cent of businesses in London did not have terrorism cover rising to seventy per cent or more in the provinces. However, to keep the terrorism pool viable, premiums needed to be increased, as the firms taking out cover were in predominantly high-risk locations. This was a problematic issue for the DTI:

> We were looking at a substantial rise in terrorism costs after the second bomb as the City more than ever before is seen as being vulnerable to a third bomb. However, there was a feeling that if Pool Re hiked-up the cost too much then there would be problems.
>
> (Insurance Analyst)

Furthermore, the chief executive of the loss assessors for the Corporation of London, warned that the City could be severely damaged financially if further terrorist incidents forced premiums too high: 'If the Government won't pick up the tab and the insurance companies won't, the City will shut down and the terrorists will have won' (cited in Pendlebury, 1993). In particular, the CBI was vociferous in their disapproval of the supposed rate rises. Their policy advisor indicated that rates 'would place an intolerable burden on hundreds of firms, particularly those in the City of London' (cited in Lapper, 1993). A BIIBA spokesperson indicated that such rates could frighten business away and decrease the City's competitiveness creating, in Beck's terms, an uninsurable 'loser region'.

Indeed, immediately after the bomb the ABI noted that the whole of the City might have to be classified a 'high risk area' and get similar risk premium ratings as 'target risk', 50% more than other Zone 1 risks:

> With two bombs in 13 months...insurance companies are likely to reclassify the whole of the financial centre as a target risk with an inevitable increase in premiums.[9]

New Zones of Risk

Revisions to the premium structure following the rate review were announced on 3 June 1993 where a number of changes were made in relation to the cost of insurance. These included the creation of four Zones (A-D) for the calculation of premiums instead of the two (Zones 1 and 2) that previously existed. This was a more realistic zoning in terms of who was potentially at greatest physical risk, and was trying to extract most money from those perceived to be at most risk. The re-alignment of zones was done as the Government decided that it must increase premiums to produce a sufficient cash-flow into the pool to achieve its stated objectives of nil cost to the Government, and to be able to withdraw from the scheme as soon as possible.

Under the new scheme, Zone A, which comprised Central London with the postcodes W1, WC1, WC2, SW1, EC1, EC2, EC3, EC4, SE1, E1, E14, had a rate of 0.144% per annum of total sum insured. This amounted, in some case to a 300% increase in premium. Zone B (rate of 0.072% - increase of 200%) comprised the rest of London and the central business districts of most other major towns. The rest of London had previously been classified as high risk in the old Zone 1. Figure 6.1 shows the geographical distribution of the new Zones throughout London. In particular it shows the concentration of Zone A postcodes around the City (EC1-EC4) and West End, indicating the most likely geographical targets for the terrorist attack based on the insurance perception of risk.

Zone C on the new scale (rate of 0.018% slight increase for smaller portfolio's but up to 80% increases on portfolios of £500 million or more) comprised the rest of England with the exception of Devon and Cornwall. Zone D (rate of 0.009% - decrease of between 10-50%) covered the rest of Great Britain i.e. Devon and Cornwall and Scotland and Wales (with the exception of the major urban areas covered by Zone B).

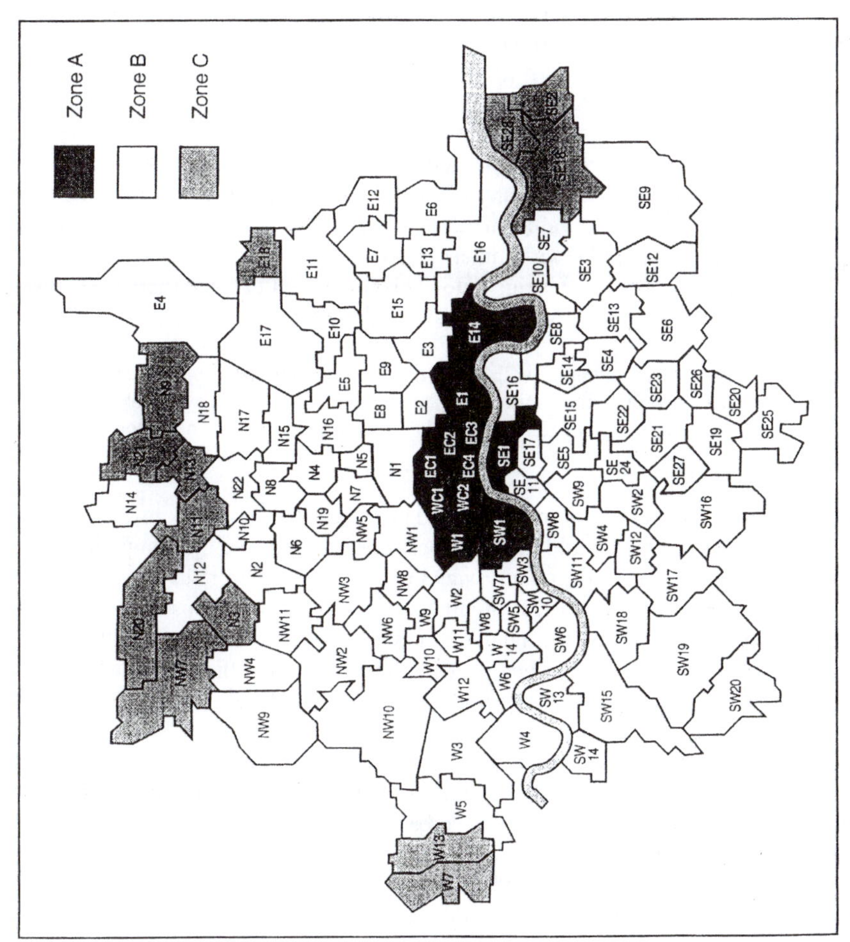

Figure 6.1 The London postcode districts used by Pool Reinsurance

Premiums in lower risk areas were decreased in order to encourage more companies to add to the terrorism insurance pool. This rating structure attempted once again to spread the risk of terrorism nation-wide and allow good risks to subsidise bad risks, although through a slightly different mechanism than before.

City Premiums Continue to Rise

City of London premiums were hit particularly hard in this rate review especially given the high property value of many of the buildings located there. For example, property in the City with a value of £500 million had a terrorism premium for material damage of £175,000 (without target risk classification) prior to 1 July 1993. This increased to £720,000 after this date. This amounted to a net increase of over 300%. Subsequently many City businesses were highly critical:

> Our clients were very unhappy with what they saw as an extortionate increase. They would not have been that bitter if the rates had doubled given the recent bomb in the City, as they would see this as realistic. But frankly 300% was a bit steep to say the least.
>
> (City Insurance Broker)

The large increases in insurance rates had potentially damaging consequences, in that it was feared that they could raise business costs and reduce the competitiveness of the City. The business community was unhappy with such rises as it felt other areas of the country were at risk from attack. For example, they used the bomb in Gateshead in the North East of England, where a gas-storage depot was attacked in June 1993, to promote this argument. As with the aftermath of the 1992 bomb, it was feared that businesses could consider relocation as a reaction to the City being given high insurance rates.

The concentration of financial risk in the City reinforced the view that the ring of steel was necessary, and that the innumerable private security initiatives being undertaken were justified. In short, increased insurance premiums meant that increased fortification was both required and initiated by the Corporation, as the insurance industry had not managed to spread the risk sufficiently to leave them financially secure if a major bomb was detonated in the City.

Fluctuating Risk Levels and Alternative Risk Distribution Mechanisms

On 31 August 1994 the Provisional IRA announced a ceasefire, which prompted immediate calls to Pool Re regarding possible premium discounts. Their Chief Executive issued a statement at this time indicating that they would not be making any immediate changes to their policy as they felt the Provisional IRA were not the only terrorist threat to the UK:

> Together with our members, Pool Re welcomes the cease-fire and very much hopes it will be permanent. At this stage it is too early to make any changes in the premium rates, but

> the progress of the peace process together with the overall claims experience of Pool Re will be kept under review. It should be pointed out that terrorism is not easily predictable and the recent attacks on the Israeli Embassy and at Finchley demonstrate that the threat is present from organisations other than the IRA.[10]

This could be seen as an attempt to maximise revenue into the terrorism pool in case the ceasefire broke down and the Government became financially liable. At this time many in the business community were not happy with what they saw as the sluggish response of the scheme to the new conditions. Some even hoped that Pool Re could be scrapped. As one insurance broker indicated: 'when the ceasefire was signed in August, I had loads of clients ringing me up and asking me if this meant they could cancel their terrorism coverage.'

Initially Pool Re indicated that rates could not be reduced. Then in December 1994 signs began to emerge that concessions for buyers of terrorism insurance might be forthcoming. As a result of the ceasefire a number of changes were made to Pool Re which became effective, in most cases, from 1 January 1995. A deposit premium of 60% of the annual terrorism premium was paid and the 40% balance would only be payable should estimated claims for acts of terrorism exceed £50 million during 1995. This was seen as both a commercial necessity, and as a response to the continuation of the peace process in Northern Ireland:

> Pool Re was under tremendous commercial pressure to reduce premiums but they were also under some political pressure to show the benefits of the progress the Government were making in working towards a total cessation of hostilities.
>
> (Senior City Insurance Consultant)

He continued by noting that the insurance industry, in his opinion, were not that keen on reducing premiums payable as it would decease the terrorism pool and heighten their possible exposure in the case of a major incident:

> The insurance industry wouldn't have been involved in the pressure [to reduce premiums] because they had walked away from the problem. In fact the insurers stood to lose. Their interest was to keep the income as high as possible due to the 10% claw back they are liable for if Pool Re funds are exhausted. One could almost see some reluctance by the insurers to see reductions in premiums.

Subsequently the cease-fire period held and no claims were made against Pool Re in 1995. The discount scheme was then extended into 1996, this time with a £75 million limit.

Renewed Vulnerability

The Docklands bomb in February 1996 meant that, once again, uncertainty surrounded the cost and provision of terrorism insurance. In addition it was also reported that some of the worst affected buildings were not covered by terrorism insurance. After this incident, Pool Re announced a considerable increase in the number of clients taking out cover with them. Considerable interest was also being shown in the alternative schemes that had entered the market.

Other Provisional IRA bomb incidents in London in the weeks preceding the South Quay blast further served to publicise the need to obtain terrorism insurance.

On 13 June Pool Re announced that the South Quay bomb had topped the threshold limit of £75 million and that they would be calling in the 40% deposit from businesses that had taken out cover. Two days later a massive lorry bomb ripped the centre of Manchester apart with damage initially estimated at between £200-300 million. This estimate later increased to £400 million.

The take-up of terrorism insurance by UK businesses increased as the perceived physical risk became more widespread. In particular, events in Manchester shattered the commonly held belief that businesses would be safe if located outside London, hence distributing the financial risk of terrorism to a greater degree than it had previously done. As a City insurer stated: 'Following South Quay the volume of business from both London and the regions increased a great deal and we are forecasting that a similar thing will now happen, particularly in other areas of the country than the capital.'[11] The combined result of these bombs had effects on two key areas of business. First, it kept terrorism on the agenda for business that continued to have to pay for it, and second, it increased the geographical scope of terrorist risk and hence encouraged those in locations previously considered at lower risk to take out coverage, and increase the amount of money flowing into Pool Re.

Available Alternatives to Pool Re

As soon as the terrorism insurance issue became a serious issue in late 1992 there were individual brokers, both in the London market and from abroad, who were initially prepared to offer cover against terrorist risk. At this time, however, the limited capacity of such schemes meant that they were relatively unimportant in the market. For example, Bain Clarkson offered a simple first loss insurance of £20 million for any one incident in May 1994. As previously noted, the huge potential losses made the Government the only viable scheme.

The reduction in terrorist activity after the Provisional IRA ceasefire in August 1994 led to the emergence of a competitive alternative market and in early 1995 the first feasible non-Government backed form of terrorism insurance became available. Such alternative schemes began to receive much more attention after the South Quay and Manchester bombs in 1996. At this time there were a large number of businesses who were reconsidering their decision not to take out cover with Pool Re. Many of these were going to the alternative schemes as well as Pool Re for a quotation. The alternative schemes operated a kind of 'cherry picking' arrangement selecting certain risks in certain areas whilst being careful not to insure too many buildings in high-risk areas because they did not want to over-expose themselves:[12]

> The alternative schemes have a specific capacity, they will insure in the City, and then, they won't take any more. We can't do this under Pool Re. This will leave us and others very exposed if the IRA come back again.
>
> (City Insurance Broker)

The alternative schemes had a number of common features. They provided a non-selection policy whereby the client only needed to insure those buildings in locations they thought were at risk. Whereas Pool Re required the entire portfolio to be covered regardless of risk perception (as this would increase the money going into the insurance pool) the alternative schemes also assessed risk management discounts on an individual basis and employed the services of risk management specialists to do this. This meant that properties seen to be more security conscious (such as through increased fortification) were given reduced premiums. In short, alternative policies were tailor-made for specific client need and, unlike Pool Re, offered a cancellation policy which many companies saw as important given the faltering nature of the peace process. Perhaps they encouraged some businesses to stay in the City, by offering them significantly lower rates than Pool Re.

The alternative schemes were seen to be in direct competition with Pool Re as they had the ability to affect the amount of money going into Pool Re and hence reduce Pool Re's ability to spread the financial risk of terrorism nationwide. As a Senior AIRMIC consultant, noted:

> I suspect that these schemes affected the way Pool Re operated because if there was no other option it would be much more rigid. With the advent of these schemes they realised money was going elsewhere.

The DTI were closely monitoring these schemes to try and assess how much non-Zone A premiums were contributing to the Pool. A further danger to Pool Re revenue also began to present itself, as larger global corporations were, in some cases, relying on their own in-house insurers rather than joining Pool Re, or taking a chance on terrorism, knowing that their company could easily absorb the financial impact of a terrorist strike.

In the latter years of the twentieth century and early twenty first century, as funds for Pool Re grew and the terrorist threat receded premium rates fell significantly and many companies declined to take out cover. Despite this Pool Re still maintained a sizable membership. At the end of 1999, Pool Re had 213 members spread between UK companies (104), Lloyd's syndicates (32) and insurance companies in the European Union and other parts of the world such as Australia and the US (77).

By the end of 2000 the company had accumulated a large surplus of £665 million. However gross premiums significantly fell from a high of £369 Million in 1994 to only £39 million at the end of 2000. This reflected a lack of coverage 'take up' as well as an 85% discount rate given by Pool Re on premiums as a result of no terrorist attacks.

Insurance and Fortified Security

The changing risk agendas of the 1990-2001 period brought about a series of changes in the way in which the insurance industry attempted to spread the financial risk of terrorism. At the same time there can be seen to have been a series of relationships developing between the way in which the insurers viewed the risk of terrorism and the way in which it was perceived by agencies of security. The processes involved can be seen to be subtly supportive of one another as both were concerned with maintaining the reputation of the City as a safe and secure business centre. The behaviour of the insurance industry can be seen to have been important in influencing processes which resulted in the increased fortification of the urban landscape, as well as educating business about the risks faced. In particular, contingency planning was widely utilised as an anti-terrorism measure.

Given the confusion over insurance cover and the risk of further terrorist attack great emphasis was put on contingency planning as a way in which the City could survive another attack and get back up and running as soon as possible if one occurred. This was important from an insurance perspective, as terrorism insurance covered business interruption costs as well as material damage.

There were two main elements to such contingency. First, there was crisis recovery planning (CRP) which highlighted how the City could bring about a 'business as usual' situation as soon as possible. Second, security plans were developed which included the risk management response such as CCTV and access control which formed the basis of protective security.

Crisis Recovery Planning

CRP was initiated at two levels; first in relation to individual companies, and second, in respect to a City-wide plan. After the St. Mary Axe bomb many companies had been preparing CRPs. In particular, such plans made contingency for temporary relocation to 'disaster recovery space' at short notice. Businesses had been preparing for another bomb ever since the St. Mary Axe explosion.

After the Bishopsgate bomb one million square foot of alternative office space was sought. Some companies even had disaster space purpose built including telecommunications links and computers, whilst others formed partnerships and leased a building for this purpose:

> Most businesses have 'hotsites', which allow them to continue to work at 20-40% capacity at a different location in the event of a disaster. Such sites have already installed IT and phone lines. Some are privately owned by the companies concerned whilst others are owned third party. There are also 'warmsites' which are less ready but still designed to cope with the initial response. These sites are often by reciprocal arrangement or through using the unoccupied office space in the City.
>
> (City Security Advisor)

At the time of the Bishopsgate bomb, CRPs instigated by the Corporation of London at a City-wide level were also in evidence. The Corporation's disaster plan had been refined through practice drills and aimed to get people back to work as quickly as possible. Michael Cassidy also noted:

> Like the first bomb, after Bishopsgate the Corporation's efforts were clean-up, shrink the police cordon around the bomb site as quickly as possible to get people back to work, and conveying a PR message that we could take it and that things would be normal.

As the British Prime Minister at this time, John Major, indicated after Bishopsgate: 'we want to show that this type of terrorism does not pay dividends. We want to get the City back and working again and show that they will not disrupt the commercial heart of the country.'[13] As Baily (1993, p.3) concluded, 'the juggernaut of the City had shuddered and slowed, but it never stopped moving'. Importantly, the international finance community, commending the quick response of the Corporation, indicated that the bomb would not drive them out of the City.

CRPs employed after Bishopsgate were, as noted in the Corporation's *City Research Project* (1995), themselves a result of corporate change at a global level. In particular: 'the large institutions which have taken the most extensive measures have not done so solely because of a specific threat [terrorism] but rather as part of a global scheme, not least to counter infrastructural failure such as power cuts and flooding (another danger in London).' The Corporation continued by indicating that in their opinion their efforts were superior to those in other financial centres: 'The level of contingency planning both by the Corporation and by individual firms is in contrast to centres such as New York, where the response to the World Trade Center blast was impressive but *ad hoc*.' Furthermore, this report highlighted that the City viewed the threat of terrorism as something to be proactive against as opposed to just reactive: 'The degree of fatalism with regard to deterring terrorist incidents observed in officials interviewed in the United States was in contrast to the proactive approach of the City after its first bomb.'

Most commentators agreed that the recovery plans used in the City were well structured, and successfully conveyed a 'business as usual' message to the outside world following the attack. As Michael Cassidy indicated during an interview:

> Over 400 businesses contacted us [the Corporation] in the month after the explosion to enquire how they could improve security and contingency planning procedures. Enquires also came from New York as to how they could improve their disaster recovery planning.

At this time property companies in the City were keen to show that confidence in the Square Mile had not been dented. However, they were forced to concede that there was still a great deal of corporate worry about the impact of further bombs and higher insurance premiums, although other concerns were also evident. As one property agent noted, 'people are undoubtedly looking to move out of the City because of the bomb, but high rents and the feeling that the grass is greener are equally valid reasons' (Jacoby, 1993).

Security Planning

Security planning like CRP at this time can be expressed on two levels: first the private response of individual companies in terms of risk management measures, and second, the co-ordinated response of the police and the Corporation of London in constructing a ring of steel and associated security infrastructure.

A key reason risk management measures are commonly undertaken in everyday life by individuals and businesses is because they are looked upon favourably by insurers. A common complaint made about Pool Re's rating system is that initially it did not provide any premium reduction incentive for companies or local authorities to take risk management measures. As previously noted, the insurers and the DTI who were running Pool Re were not keen to give incentives to business in the early stages of the scheme as they wanted to maximise the money going into it.

According to the City of London Police, they tried to talk to the insurance companies independently, and to the ABI, to see if they were willing to offer an incentive to businesses in the City to put up cameras for crime as well as terrorism purposes. At this time the insurers were unwilling to offer such incentives. The attempts by the Police to work with the insurance industry were hardly surprising. They had been severely criticised both publicly and privately for failing to stop the second major bomb and were keen to do all they could to stop a third incident occurring:

> It was me, or us [the City Police], that were trying to push insurance companies into offering a discounted premium because we thought that it would encourage greater security measures to be implemented or installed...but it never came to pass.
>
> (Senior City Police Officer)

However, the insurance industry contributed to the reinforcement of security in the Square Mile through policy changes in the Pool Re underwriting manual that increasingly gave the opportunities for premium discounts for occupiers who increasingly fortified their buildings through use of risk management measures.

The review of Pool Re in July 1993 introduced improved discounts for risk management measures for up to 12.5% of the total policy premium. Despite initial problems, most insurers, in time, complemented the risk management scheme[14] as it became far more user friendly as not all buildings in a portfolio needed to comply to obtain a reduction. Subsequently renewed, and more workable, risk management incentives began to be offered to policy holders from June 1994 after considerable lobbying from the insurance industry. As a Corporation of London's insurance officer, indicated during an interview: 'I'm pleased a system of risk management discounts came in. We are now talking about big money. We are pushing it.'

For example, the Corporation had to insure around £5-6 billion of property so any risk management discount was welcomed. The Corporation of London were certainly very aware of possible discounts that local businesses could claim. As noted in Chapter 5, after the Bishopsgate bomb the Corporation of London employed a specialist security advisor whose job encompassed liaison with City businesses in terms of how they could improve security and reduce business interruption losses in an event of a terrorist incident. As indicated by Michael Cassidy:

> His job is to go around City businesses encouraging them to come in on the Camera scheme [CameraWatch], to have contingency plans, take out the [insurance] cover and all the rest of it...Part of his pitch is that you can get discounts for insurance.

As well as individual firms and organisations improving risk management with the hope of improving their security and of getting a discount from Pool Re, a co-ordinated security response was organised by the police and the Corporation of London in the form of the ring of steel and associated measures.

At an official level, the insurance industry did not play any part in the actual decision to construct the ring of steel in July 1993. This was a decision taken by the Corporation in conjunction with the City Police to help retain international confidence in the City as a good place to trade. As Michael Cassidy indicated, 'there was no direct pressure whatsoever from the insurance industry for such measures'. A senior City Insurance broker also noted:

> The insurers didn't really worry as they had washed their hands of the situation and could not find themselves in financial straits because the Government were backing up the scheme.

Whilst the ring of steel, in the opinion of most, enhanced security, these risk management measures provided the most concrete example of a proactive security strategy which was *unable* to elicit a financial discount from Pool Re. A senior City Insurance Consultant indicated that the City, after Bishopsgate, wanted a premium reduction for the ring of steel:

> Among the problems with Pool Re at this time was the failure to give premium reductions for risk management measures taken. The Corporation of London, with their road blocks and camera network, were a particularly obvious example. They were all aggrieved that the City were not getting discounts for their exhaustive efforts to thwart the terrorist threat.

Others expressed the view that the substantial reductions in general crime in the City (10.6% in 1992 and a further 17% in 1993), attributed to the ring of steel, should have elicited a more favourable response from insurers regardless of terrorist implications.

This relationship between insurance and physical fortification in the Square Mile is complex, and views were mixed as to whether businesses inside the cordon should have received a discounted premium. This, of course, would have led to a reduction in the money being collected by Pool Re and increased the liability for both the Government and insurance industry.

There was also a strong suspicion that if the ring of steel were removed there would have been significant problems of insuring against terrorism within the Square Mile. In this scenario premiums would have further increased according to some respondents:

> Of course the premiums would increase if it [the ring of steel] were removed, however people in the City are not prepared to see it relaxed irrespective of insurance cost.
>
> (Corporation of London Insurance Officer)

This comment indicates that insurers and the Government saw the ring of steel as an effective risk management strategy given that they were happy to insure in the area if the ring of steel was still active. Underwriters of terrorism insurance also took a lot of comfort from the ring of steel. As one Insurance broker indicated, the security cordon was of benefit in this regard as: 'it gives far greater security to the insured risks located here.' Another senior City insurer, also noted that those in the insurance industry felt much safer assessing policies with the security cordon in place: 'underwriters have taken a lot of comfort from the ring of steel.' Furthermore Michael Cassidy noted:

> I think insurers would be alarmed if we didn't have it. I think given our special history it would be stretching it to expect any kind of discount. If the ring of steel were removed I think there might be problems of obtaining cover.

He continued by portraying a more realistic view of the situation, indicating that since the City was the number one Provisional IRA target, and that the ring of steel could perhaps increase the risk, it would be illogical for Pool Re to give a discount for the ring of steel.

> I don't think a premium reduction for the ring of steel is possible... because the City is the prime IRA target and you can't guarantee it [the ring of steel] isn't going to be breached. It could certainly be breached with small hand-carried bombs and it was even argued at the time that having the ring of steel is an incitement to terrorists.

In addition:

> From this point of view I think it would be [have been] unreasonable to turn around to the insurers and say 'we are so safe here – are you going to give us a discount'?

This opinion was reiterated by a City Insurance broker who also noted that the ring of steel could help to indirectly lessen premium costs by restricting the potential for terrorist attack in the City. This would feed back into reduce premiums:

> A discount for the ring of steel is a chicken and egg situation. If the ring of steel has a positive effect and the City avoids future losses then the absence of losses will feed through to the terms and conditions of the facility. However the problem remains that you can't guarantee 100% security.

This, in short, meant that whilst the City could not obtain a discount for the ring of steel, its removal would have could caused a potential crisis in the market with regard to the provision of insurance cover.

Globalisation of Risk

The events of 9/11 were by far the biggest and most complex case the insurance industry has ever had to deal with. The magnitude of this impact will take years to unravel fully and will have significant impacts for insurance and reinsurance markets and the business community in general.

First estimates of potential losses including physical damage, loss of business, and workers' compensation, life and disability claims, was between $10-15 billion with Lloyds of London and the larger worldwide reinsurers indicating that losses would be 'painful but manageable'.[15] Others suggested the claim would be much higher at between $40-70 billion. Before 9/11 insurers in America included terrorism as part of standard coverage when they claimed the risk of potential losses was small.

In the aftermath of 9/11 the US government then set about trying to find a way to protect insurance companies from further claims caused by terrorism. Immediately the UK Pool Re model was put forward as a starting point for any US model as reinsurers hinted they would raise rates and would not offer coverage against terrorism as they feared for the industry's solvency.[16]

Comments made by the insurance industry in America post 9/11 echo those almost a decade earlier in the City of London. For example The Chairman of the American Insurance Association (AIA) noted that:

> It's not an issue of profitability for the industry; it's an issue of economic stability to the entire economy. Without insurance protection from terrorists, banks will hesitate to approve loans for real estate, construction and manufacturing.[17]

He continued: 'No one can price this risk. It's not an act of God; it's an act of man designed to inflict maximum damage and destruction.' The president of the Reinsurance Association of America further noted that 'one could assume a complete reassessment of catastrophic risk is underway'. In short it was highlighted that terrorism could not be modelled or underwritten like other catastrophic events, as it is not susceptible to the normal 'laws of insurance'. Acts of global terrorism, it was noted:

> Exposed the insurance industry to infinite risk; as unlike natural catastrophes which can be 'zoned', have historical-temporal patterns, and are random, terrorism is not random (sites are chosen), and terrorists deliberately use unpredictable patterns to evade capture meaning such zoning is difficult with any degree of accuracy.

As such a Pool Re type solution was called for:

> If the federal Government doesn't come up with a proposal like this for the private sector, who's left holding the bag? It's not the reinsurers, it's the primary companies. What do you do? You just stop writing the coverage. You can't price it. And a primary insurer can't absorb the risk.
>
> (President of the Reinsurance Association of America)

He continued by highlighting its impact on the general state of the economy: 'The alternative is insurance marketplace chaos, which means a ripple effect in the economy could be quite significant.'

Subsequently, the insurance industry put forward a suggested model called Homeland Security Mutual Reinsurance Company, based on Pool Re, as part of the 'Insurance Stabilization and Availability Act of 2001' which was formally proposed on 10 October. Five days later the Bush Administration published their own 'terrorism insurance plan', which proposed that the insurers picked up a great amount of the risk up to a certain level ($100 billion). The insurers would be responsible for 20% of the first $20 billion losses in year 1; 100% of the first $10 billion, 50% of losses between $10-20 billion and 10% of the losses above $20 billion in year 2; and 100% of the first $20 billon in year 3, and so on. A further scheme was mooted on 1 November 2001 – The Terrorism Risk Protection Act which would be triggered only in the event of a $100 million loss. The federal government would, it was proposed, then pick up 90% of the loss leaving the insurance companies to pick up the remaining 10%.

These three main proposals were debated for over a year until eventually in November 2002 the Terrorism Insurance Bill was signed which in essence was a compromise between the solutions mentioned above. The act was designed to facilitate the development of a robust market for terrorism insurance. It insisted that insurance companies offered terrorism coverage but also allowed for a transitionary period for markets to stabilise and build the necessary capacity to sustain any future losses. As such the federal government would pay 90% of all losses until the end of 2005.

The UK terrorism insurance market was also reappraising the situation post 9/11. In December 2001, the UK Government began discussions with the insurance industry to broaden arrangements for terrorism insurance given the transformed perceptions and understandings of global terrorism post 9/11, and the changing nature of the insurance market since Pool Re was established in 1993:

> The tragic events of September 11th have forced everyone to re-examine the risks we face from modern-day terrorism. The insurance industry is working hard to ensure that as much cover can be provided as possible against exposures that can be enormous. Pool Re is an outstanding example of how Government can work with the private sector to the advantage of the whole community. We should now build upon its achievements.
>
> (Director General of the ABI)[18]

The totally catastrophic events of 9/11 were further highlighted at the end of the financial year (2001-2002) when the London insurance market announced losses of over £3 billion with nearly £2 billion coming directly from the 9/11 attacks. However it was also mentioned that the events on 9/11 had helped restructure the insurance market. As an insurance agent noted:[19]

> The fact that the World Trade Center has acted as a catalyst for significant improvements in ratings and terms and conditions means Lloyds at least can look forward to the prospect of a decent profit with genuine optimism.

In short the insurance industry had been in difficulty for some time and had been experiencing large underwriting losses. Even prior to 9/11 it was widely estimated that insurers were paying out £1.15 in claims for every £1 received in premiums. As such some commentators have accused the insurance industry of using 9/11 'as an excuse to raise premiums across the board to make up for falling profits in recent years' (Madslien, 2002).[20] The argument is that the restructuring and streamlining of the insurance industry has been speeded up by the tragic events of 9/11 by forcing the insurance industry to be increasingly reflexive in their thinking and introduce cost cutting and efficiency improvements. As such the fallout of 9/11 has seen premiums in many forms of insurance rise and job losses announced at some companies. For example, in September 2002, Europe's third biggest insurer, Zurich Financial, announced around 4500 job losses as a result of $2 billion losses in the previous six months, in large part attributed to the impact of 9/11.

The new structure of Pool Re announced in July 2002 increased the definition of terrorism to which previously had covered fire and explosion only. The new broader definition, covered 'all risks', including contamination by biological, chemical or nuclear (after January 2003) agents, the use of aircrafts in attacks or, flood damage. As a result of these changes premiums were doubled until the end of 2002. After this date insurers were able to decide premium rates on a 'case by case basis' according to normal commercial arrangements. This introduced a greater degree of flexibility and competition in to the market and encouraged the provision of terrorism insurance from private sector reinsurers. In addition each insurer will have its losses capped per event and per annum so insurers will know in advance the maximum they might have to pay out. Initially from January 2003 this was £30 million per event or £60 million per annum and will gradually increase over the years. This arrangement will allow the reinsurance market time to re-establish its financial capacity in the market. Table 6.1 below highlights the expected increase in maximum industry retentions.

Table 6.1 Maximum losses faced by the insurance industry for acts of terrorism

Start Date	Maximum retention per event (£m)	Maximum retention per annum (£m)
1 January 2003	30	60
1 January 2004	50	100
1 January 2005	75	150
1 January 2006	100	200

Given the distribution of claims within the insurance industry it is unlikely this maximum retention would be exceeded.

Conclusion

Since 1990, economic and political processes operating at a variety of spatial scales shaped the financial distribution of terrorist risk in relation to the City. On a global scale, international reinsurers dictated that the UK direct insurers should withdraw from the market, hence creating a situation where the City's pre-eminence in the global market place was questioned. The refusal, or inability, of UK insurers to underwrite terrorist risk meant that the national Government was forced to act as reinsurer of last resort. The Government subsequently attempted to redistribute its possible financial losses from terrorism throughout the national economy through the original Pool Re scheme, which attempted to generate a large amount of money from around the country to cover the potential losses of a few high exposures areas such as the City. Whilst, in theory, this scheme should have worked well, the more realistic approach taken by owners of commercial property as to whether or not they were actually at risk from terrorism (given that the Provisional IRA were mainly targeting the areas of economic importance in London), meant that in reality risk distribution was defeated, and the financial risk became concentrated in the City. In time, significant adjustments were made to Pool Re (and alternative schemes developed), which sought to once again to spread the risk more equitably across the country.

Within the aforementioned processes a key voice has been the Corporation of London who throughout this period constantly attempted to persuade the Government to adopt insurance mechanisms which would decrease the insurance costs to occupiers in the Square Mile, and give greater confidence to those businesses seeking to locate in the City. To this extent the Corporation had mixed fortunes. On the one hand they were influential in getting the Government to financially back up terrorism insurance, although the insurers would have preferred a Northern Ireland type scheme, which left them with no liability. On the other hand, the Corporation were less successful in persuading the DTI to reduce City premiums. Rises of up to 300% were experienced in June 1993 which, perhaps, accurately reflected the reality of risk, but was also an attempt by the Government to increase the amount of money entering the Pool Re scheme, given the relative lack of interest from outside London. Thus the Corporation attempted to gain insurance discounts, whilst at the same time constructing a vast array of physical fortification measures to reduce the threat of attack. The next chapter will fully explore the role of the Corporation in attempting to deal with the risk of terrorism in the City.

However, the fact that some form of insurance could be obtained against terrorist risk undoubtedly helped the competitive position of the City at this time. Pool Re in this sense achieved a great deal. It enabled the City to continue to trade secure in the knowledge that the cost of damage resulting from terrorist bombs would be met through insurance, and allowing the City (as well as London's role as a financial centre, and the UK economy as a whole, to grow as the twenty-first century approached.

Importantly, it was the links between the provision of insurance cover and fortification measures that underpinned this success. Risk management incentives for which insurance discounts could be obtained had a pronounced spatial effect on the City landscape. More individual City properties produced adequate contingency plans and increasingly fortified themselves against possible terrorist attack, and in so doing reinforced the security effort in the Square Mile. This occurred through a proliferation of external CCTV, and highly visible security guards who operate computerised access procedures. Internally, many building occupiers made extensive use of CCTV, blast resistant curtains and anti-shatter window film. At a City wide level the ring of steel and the centralised security schemes and response plans put in place by the City Police were seen as essential if the City occupiers were to be able to obtain affordable terrorism insurance.

As with physical security, the events of 9/11 have led to significant reappraisal of the insurance mechanism for combating terrorism both in the UK as well as noticeably in the US. The enormity of the losses experienced by insurers around the globe on and after 9/11 has led to the development of new and refined schemes to offer coverage against terrorist attack as well as leading to a host of cost cutting and efficiency drives by insurance companies as they begin to realise the enormity of potential losses faced through acts of terrorism.

Notes

1 Companies who reinsure part of their risk with reinsurance companies.
2 On 12 November the *Financial Times* ran an article based on leaked material from AIRMIC indicating the reluctance of reinsurers to cover terrorism. This it is believed triggered an ABI press statement confirming this state of affairs (AIRMIC Newsletter January 1993).
3 ABI press release, 12 November.
4 The City Property Association represented a wide range of membership, including the major financial institutions, legal firms and property investment companies, as owners and occupiers of property within the City of London.
5 Cited by Peter Sharp (reporter) on *News at Ten* (ITV), 18 December.
6 Cited on the *Nine O'Clock News* (BBC1), 21 December.
7 Cited by Kynoch and Rydell (1993).
8 For example Sun Alliance insured a large proportion of the Corporation of London buildings, approximately one-third of the property in the City.
9 A leading member of the ABI cited – *New Builder*, 30 April 1993.
10 August 1994 saw two separate terrorist attacks in London, one at the Israeli Embassy. Pool Re were quick to confirm that, despite these attacks not being aimed at the UK Government, their reinsurance cover would respond to the losses.
11 Cited in Guy (1996).
12 Of all the alternatives on the market three warrant a special mention here as they attracted widespread attention from City businesses: the BIIBA Hiscoxs and Minets schemes.
13 Cited in *The European*, 29 April 1993, p.14.
14 Much interest was shown in the potential of obtaining discounts for risk management measures i.e. taking positive measures against terrorist attack. Companies were invited to complete a questionnaire covering a number of different aspects of security from physical protection to recovery planning. Then if ALL the premises owned by the company in the UK conformed to the entire requirement a 12.5% discount could potentially be obtained on the material damage premium with a smaller discount carrying through to business interruption. In January 1995 Pool Re confirmed that it was looking at further ways to improve the Risk Management Discount system.
15 Cited in *Electronic Telegraph* – 20 September 2001.
16 Terrorism insurance models which operated in South Africa and Spain were also looked at.
17 Cited in van Aartijk Jr, 2001.
18 Cited in *Property Forum News*, 23 January 2002.
19 Cited in Cave, 2002.
20 BBC News Online, 2 September 2002.

Chapter 7

Framing, Legitimating and Negotiating the City's Response to Terrorist Risk

Introduction

The previous two chapters have highlighted how the nature of terrorist risk in the City was addressed differently by the agencies of security and the insurance industry. In both cases the key governance role of the Corporation of London was highlighted as critical in allowing the ring of steel to be established and in putting pressure on the Government to set up a terrorism insurance scheme. As such, during the 1990s the Corporation of London became the principal driving force behind defending the City from both the physical and financial risk of terrorism. Previous chapters have also noted how the Corporation attempted to re-articulate their response to terrorist risk in terms of crime reduction, better traffic management and environmental improvements, and how the importance of the City in the national and London-wide context was pushed to the fore as justification for the creation of the ring of steel. This chapter will investigate these assumptions showing how the Corporations response to terrorism was framed, legitimated and negotiated out as a result of institutional alliances between themselves and other key urban managers and political authorities, creating a powerful 'pro-security' inside discourse.

However, it will also be argued that the inside discourse, despite capturing a general feeling that security should be enhanced, was not truly representative of the variety of views expressed by the different interest groups within the City. This chapter will show that there was not a homogenous 'community of interest' around enhancing security, but rather a number of different spatial interests whose views were reflected in the way the ring of steel was developed. It will be argued that such views were constantly in a state of flux and were modified in relation to prevailing socio-economic and political conditions.

This chapter will also detail the opposition to the ring of steel, which formed the 'outside discourse'. This came from a number of institutions and organisations, most notably civil liberty groups and neighbouring borough councils. Furthermore, it will be argued that these views were marginalised and powerless in altering the strategies of the Corporation.

This chapter is divided into six parts. The first part will briefly note the historically and geographically specific institutional arrangements within the City, showing how these framed the construction of a powerful 'inside discourse'. It will also be shown that the views of the inside discourse were, in part, also constructed by the way terrorist risk in the City was being portrayed in the media, and also by the concerns expressed by central Government that terrorism could affect the portrayal of London as a 'world city'. Second, this chapter will show how the Corporation of London sought legitimacy for their security proposals through the consultation process. The third part of this chapter will argue that as a result of consultation it became obvious that whilst the need for security was accepted, there were strong disagreements between different interested parties within the City as to what form and duration such risk-management measures should take. Fourth, this chapter will investigate how the dominant views expressed by the powerful economic and political elite were contested by groupings operating outside the institutional networks of the City. Fifth, the chapter will address how the City, as a result of terrorist risk, successfully operationalised local institutional strategies which strengthened the position of the Square Mile within the global economy. Finally the sixth part will highlight the impact of 9/11 on managing the security risk in the City.

Institutional Thickness in the City

The growth of institutional thickness refers to a gradual process of change within urban governance that has taken place within the context of particular local histories and organisational realms which are subject to a specific set of pressures from wider economic driving forces. To properly evaluate the changes in institutional arrangements that were established in the 1990s the 'traditional' institutionalism in the City must first be noted.

The historical development of a unique type of institutional culture within the Square Mile has been fundamental to the success of the City as a financial centre for many years. The 'traditional city', it has been argued, had a number of key features, which allowed the Square Mile to reproduce itself and remain globally competitive. Of particular note is its relationship with the British state which has been one of 'limited interference', with the job of running the City given to the Corporation of London. Furthermore, the development of a distinctive social structure based on face-to-face contacts and a series of interconnected business networks meant that spatial concentration of the Square Mile was maintained 'keeping the City in the City' (Thrift, 1994a).

In the post-war years the City was forced to partially reinvent itself as a result of the growing international financial system, which meant change was needed to transform itself from an international city to a global city with appropriate institutional arrangements, deregulated markets, and the construction of buildings suitable for global finance (Pryke, 1991; McDowell, 1994; Crang, 1998). The 'de-traditional City' therefore emerged from the mid 1980s, built around new institutional agendas – the development of trust, information sharing, knowledge production for enhancement of global economic functions and the maintenance and construction of new forms of socio-economic networking (see for example Amin and Thrift 1995). In particular, this 'de-traditionalisation' necessitated restructuring of local governance to increasingly couple local priorities and global agendas, and to enhance and maintain the City's 'institutional thickness'.

The Power of the Corporation

The strong institutionalism in the City has traditionally led to the Corporation of London being seen as a very powerful organisation within the governance of London. The Corporation is a relic dating back to medieval times and is unique as far as governance in London and Britain is concerned, having resisted all local Government reforms to date. As such it retains the institutional arrangements common in this period such as a Lord Mayor, Sheriffs, Aldermen, its own police force and a Common Council in the Guildhall. Unlike other local authorities the Corporation is non-party political (although the City is inherently associated with Conservatism) with its members elected from the business community and the major City institutions.

Historically the City became the world's leading centre of finance due to the controlling influence of the British Empire and has remained pre-eminent ever since. The Corporation's electoral role is based on businesses that reside in the Square Mile, hence powerful business voices have always dominated the local political agenda. In short, the City is run by business people to serve the need of business and has continuously resisted any political reforms which might affect its boundaries, electoral process and function (Travers *et al*, 1991).

However, in the mid-late 1980s the liberalisation of financial markets made other cities increasingly competitive in relation to the City, whilst at the same time the established institutions in the Square Mile, most notably the Bank of England, were coming in for criticism for not providing strong leadership, for losing vast amounts of money (as best exemplified by Lloyds) and, in particular, for not 'moving with the times'.

Prior to the 1990s, the City traditionally preferred a liberal culture with minimal Government interference, which was seen as one of its strengths. However, the growing pace of change in the global economy meant that this became a weakness as cities such as Paris, Tokyo and New York embraced change with new forms of quasi-Government leadership, strengthening their economic position and 'world city' status as a result. As such in the 1990s the Corporation of London was quick to initiate a series of research programmes to look at London's competitiveness in the world of international finance, in particular how the City could contribute to enhancing London's inward investment profile.[1] Promoting London's competitive position was also a key priority of central Government at this time with 'the idea of the world city permeated the discourse about London and played an important role in attempting to generate consensus over priorities' (Newman and Thornley, 1997, p.979). Furthermore London was considered at a disadvantage for 'not having a clear 'voice' to co-ordinate the promotional effort and provide a future 'vision' for the city' (ibid. p.977).

Therefore, in the late 1980s the Corporation, which had previously concerned itself with matters of financial interest, and hence remained relatively detached from the rest of London, began to widen its institutional involvement in the running of London *per se*. This coincided with a considerable entrepreneurial involvement in the governance of London as the private sector acquired a much stronger role in planning decisions. As Travers and Jones (1997, p.14) noted, this was a fundamental change for the City:

> Traditionally the City had kept a low political profile, restricting itself to representing the interests of the financial services industry and providing ceremonial colour to the London scene. But during the years since 1992, the Corporation – and most particularly the chairman of its Policy and Resources Committee, Michael Cassidy – became directly involved in a number of London-wide initiatives and partnerships...such a high profile for the City would have been unthinkable until the early 1990s.

The City's involvement at this time included the joint establishment of an inward investment agency for London – London First Centre (1993/4), alongside the City of Westminster and the London Docklands Development Corporation, which subsequently took a lead in setting up the London Pride Partnership (1994), which set out a 15-20 year vision for the development of London based on the twin concerns of economic competitiveness and social cohesion.[2] This latter partnership brought the City together with significant players from the business sector, the London boroughs and voluntary organisations. Such an approach was needed to develop a coherent and co-ordinated strategy, essential due to the fragmented nature of local government in London, which, since the abolition of the Greater London Council (GLC) in 1986, had seen the thirty-two boroughs and the City of London responsible for local government policy and programmes.[3]

Indeed, Newman and Thornley (1997, p.978) point out that the Corporation was (and still is) an active member of most London-wide boards and that 'since abolition of the GLC, the City has come to play a wider regional role and [is] far from being just a Government anachronism'. Of particular concern to the City in the early 1990s was transport infrastructure in central London, which was having ever greater demands placed on it because of the vast increase in office stock in the Square Mile.[4]

From the perspective of this book, the strong, but changing, local and London-wide institutional networks that the City was operating within in the early 1990s became very evident in the aftermath of the terrorist bombings. The 'spirit of the City' was very important in getting the Square Mile up and running again as soon as possible and in mobilising support for radical security measures to deter further attack. It also became evident that the need to respond to the actual occurrence, and the wider implications, of terrorist attack would require new forms of institutional arrangements to be constructed at a variety of geographical levels: locally through negotiation with the key economic institutions in the Square Mile with regard to the best way to respond; at a London-wide level as the City of London had to liaise with its neighbouring authorities; and at a local-national level as central Government became involved to protect what it saw as a national asset and defend the global image of London. This mirrored two aspects of London governance at the time - first, the increasingly role of central Government in London affairs, and second, increased co-operation between the boroughs.

As a result of the terrorist threat, new institutional pillars were established which effectively quickened the de-traditionalising processes the City was already experiencing. These new 'pillars' allowed the City to enhance its institutional thickness and maintain its historic power whilst increasing its economic competitiveness. The anti-terrorist proposals that were emanating from these new City-based networks also served to affect the areas directly adjacent to the territorial boundaries of the City. Before 1994 these areas fell under the jurisdiction of a number of London boroughs that were collectively represented by either the Conservative-led London Boroughs Association (LBA) or the predominantly Labour-controlled councils – the Association of London Authorities (ALA).[5] The latter of these groups is of particular importance given that the Labour party in the early 1990s was committed to abolishing the Corporation of London, which it saw as being undemocratically elected. During the late 1980s and early 1990s the ALA also had an informal policy of non-co-operation with non-Labour boroughs (see for example Travers *et al*, 1991 and Hebbert, 1999). As will be noted, the objections that a number of boroughs bordering the City had with regard to the ring of steel were expressed through the ALA, creating a climate where the City was accused of once again acting paternalistically without consideration of London-wide agendas.

The Pro-security Discourse

As a result of the continual threat from terrorism in the early 1990s a pro-security discourse dominated the City's anti-terrorist strategy. Such a strategy was developed, in large part, by a partnership involving the Corporation of London, the main businesses within the City – the so-called 'City Fathers' such as the Bank of England, the Stock Exchange and Lloyds – the insurance industry and the police. The 'collective concern' of these institutions was to maintain the reputation of the City within the global economy at all costs.

Security enhancement was seen as especially important given the negative media reporting of the recent terrorist attacks against the City. This was particularly evident after the Bishopsgate blast where media reports focused on four main themes.

First, the impact of the bomb on the City's competitiveness and on the determination of the City to maintain a 'business as usual approach' – after the Bishopsgate bomb the initial media reaction was mainly threefold. First, it highlighted the tremendous damage caused by the destructive nature of the bomb e.g. 'Cityscape of destruction' (*Daily Telegraph*, 26 April) and 'Bishopsgate destroyer' (*Daily Telegraph*, 27 April). This demonstrated the tremendous power of modern bombs, indicating that the response should be at a 'community level', rather than *ad hoc*. Second, it showed that the bomb could have a damaging affect on the City's reputation, jeopardising the City's international competitiveness, for example – 'City's reputation around the world put at risk' (*The Independent*, 25 April). Third, initial media reaction highlighted the determination of the City to carry on business as usual, sending a clear message to the Provisional IRA that they would not be defeated e.g. 'City blooded but not unbowed' (*Sunday Times*, 25 April).

Second, the media highlighted the failure of the security apparatus to prevent this attack, serving to heighten the perception of risk within the Square Mile that another bombing could occur – for example, 'It's too easy for the IRA' (*Daily Mail*, 26 April), 'IRA exploited reduction in spot security checks' (*Financial Times*, 26 April) and 'Increased security failing to combat terror campaign' (*The Independent*, 26 April).

Third, as noted in Chapter 7, the propaganda value of the Bishopsgate bomb to the Provisional IRA was further enhanced by media reports highlighting inaccurate insurable losses – for example, 'The £1 billion bomb' (*Sunday Times*, 25 April), 'Insurance Wipe-out' (*Daily Mail*, 26 April) and 'Counting the cost in cash and confidence' (*London Evening Standard*, 26 April) and the fact that insurance premiums would increase as a result – 'Insurance premiums set to soar after City bomb' (*Daily Telegraph*, 26 April).

Fourth, the media were vociferous about the need to enhance security in the City by constructing a fortified landscape, similar to that employed in Belfast, to repel further attack – for example, '"Walled City" mooted to thwart terrorists' (*The Times*, 27 April); by creating a Belfast-esque environment – 'Bishopsgate looks to Belfast' (*Financial Times*, 30 April); by constructing a panopticon of surveillance – 'Camera blitz to thwart IRA bombers' (*Sunday Times*, 2 May); or a landscape of bunker architecture – 'Bombings? We're off to the bunker' (*The Independent*, 5 May).

These headlines were again a response to what was seen as the failings on the security procedures put in place after the 1992 bomb. As Dillon (1996, p.129) noted: 'much of the media coverage of the bombing realised the IRA's twofold objective of striking at the heart of the financial centre of the capital and generating paranoia about the inability of the security apparatus to combat terrorism'. In summary, the media highlighted the need for a proactive security response from the City.

The pressure for high levels of security was maintained by continual threats from the Provisional IRA, who claimed that higher levels of security would be ineffective in stopping them bombing the area again. They attempted to create a climate of fear in the City to keep the Northern Ireland question at the top of the political agenda and to destabilise the UK economy. A statement released in Dublin by the Provisional IRA after Bishopsgate vowed to breach any security measures the City could mount:

> These latest attacks underline both the ability and the determination of our volunteers to breach whatever level of security the British Authorities are capable of mounting.

Furthermore, threatening letters sent by the Provisional IRA and received by more than fifty foreign banks and businesses in the City after Bishopsgate, warned them that they were still very much at risk.[6] This was especially true of Japanese banks that were thought to be most nervous about the impacts of terrorism. For example, a letter in August 1993 to City institutions indicated that:

> We do not seek to target those with whom we have no quarrel but the reality is that simply by virtue of their location many businesses will suffer the effects of our operations. In the context of present political realities further attacks on the City of London are inevitable. This we feel we are bound to convey to you directly to allow you to make fully informed decisions.[7]

The driving agenda of the pro-security discourse was undoubtedly to protect the economic reputation of the City. This was summed up by the Commissioner of Police who noted: 'No one should be in any doubt that we are locked into a struggle with terrorists for the City of London and it is a struggle that we, the nation (not just the City of London), cannot afford to lose.' He continued:

> I know that I need not remind you that another massive bomb could make the City untenable as an international financial market place. Foreign investments and business would flee, perhaps never to return. The £18bn a year earnings from City business could be lost and irreparable damage done to the country's economy.[8]

The Dominant Voice of the Corporation of London

As highlighted in Chapter 3, it is the dominant institutions and in particular the ruling political authority that often dictates planning and development agendas in a locality. This was exemplified after the first bomb when the Corporation were responsible for driving the anti-terrorist security agenda forward. The Corporation's response was drawn from a number of quarters, involving, in particular its leader, Michael Cassidy (who, as noted earlier was centrally involved in the marketing of London overall at this time), the City Engineer who drew up the plans for radical security enhancement, the City Planning Officer who was concerned with how the security-based plans fitted into the City of London local plan and the Unitary Development Plan (UDP), a variety of financial officers who dealt with the insurance implications of terrorist as well as the cost of implementing the security operation, and the Commissioner of the City Police who was responsible for security enhancement.

Directly after the Bishopsgate bomb the Corporation were quick to contact all the major UK and foreign institutions in the Square Mile, as well as ambassadors from around the globe, to reassure them that the Corporation was doing all it could to get the City back to normality as quickly as possible and to prevent further attacks. Initially, the Corporation had a series of meetings with central Government and organised a series public gatherings to obtain the opinions of businesses on security options. The City also instigated an informal consultation exercise with 200 prominent City businesses that were asked to comment on the 'menu' of security options. As Michael Cassidy noted, dealing with the risk of terrorism was very much in the Corporation's hands:

> In the days immediately following [the bomb] we went to Downing Street to talk about what needed to be done and what the City police thought they could do. The Prime Minister basically said 'go away and do what you can do within your own powers but don't trouble me with legislation'.

He continued:

> The Lord Mayor and I then instituted a round of contacts with businesses in the City to sound out opinion. We spoke to about 4000 over the succeeding weeks either directly, by word of mouth or by presentation. I held a series of meetings at the Guildhall and I set out some options. The idea was to test opinion as to the acceptability of radical measures.

As a result of such meetings central Government indicated that the State would be prepared to bear some of the cost of implementing fortification measures. This was further advanced by a number of additional meetings between City officials and senior Government ministers where the national and international implications of further bombings against the City were made clear. Thus the minimal Government interference agenda, which had for so long been a cornerstone of the City's success, was forced to change by the terrorism agenda, with new forms of central-local Government relationships emerging.

At a City-wide level there was a behind-the-scenes campaign, from some UK and foreign institutions, for radical security measures to be implemented. This led to substantial pressure building up on the Corporation of London to alter their security procedures to provide a more formidable deterrent against terrorism. As Michael Cassidy commented at the time: 'They made it clear that they wanted to see something happening on the streets, not just talk of improvements in the gathering of intelligence'.[9] Essentially, if the ratepayers (essentially financial institutions) demanded higher levels of security, the Corporation had to consider it. The City Police indicated that both UK and especially foreign businesses were very worried about a third bomb, and were mobilising support for a full security cordon:

> Over time we became aware of people's nervousness about a succession of bombs, in particular the Japanese banking community wrote to the Government with an absolutely classic letter, very un-Japanese of them, and said 'whilst our citizens are working in your country we expect you to look after their personal safety and if you can't then we will have to look elsewhere'.
>
> (Senior City of London Police officer)

He continued:

> It was a very direct letter. Now, you have got to respond to that and I remember that some City Fathers and chairmen of large companies were very strong that we had to take the most extreme measures that were available.

A Financial Times/MORI survey conducted in the week following the Bishopsgate blast gave an initial indication of the views of City occupiers, many of whom favoured higher levels of security. This survey highlighted that 84% of City organisations wanted an increased police presence on the street, 79% wanted the police to have improved powers to construct roadblocks as deemed necessary and 74% wanted 'formal co-ordination of security arrangements by organisations in the City'.[10] This latter concern indicated that occupiers saw the countering of the terrorist threat as a collective responsibility.

The Corporation's initial response was to employ a specialist security advisor to advise City businesses on what they could do to reduce the threat to themselves, and to organise a Crisis Response Team (CRT) on which representatives from the police, planning and financial associations could sit to discuss security issues. The Court of Common Council (23 June) noted that the formation of the CRT was 'to draw up proposals to improve the security of the City from terrorist attack while retaining the character of the City as an enjoyable place to live and work'. Therefore, as a result of the terrorist threat, both defence and environmental improvement agendas were accelerated as increased security was sought alongside restricted traffic flows, pedestrianisation and the growth of the City's residential population. It was thus readily apparent that the high levels of fortified security (like Belfast), suggested by the media would not be an appropriate response.

Subsequently the Corporation, in partnership with the City Police force and central Government Ministers, including the Minister for Transport in London, responded quickly by constructing the ring of steel in July 1993. The Government was quite happy for the Corporation to introduce measures in the short term as long as measures were experimental, meaning they would have to seek legislative permanence within a year.

Immediately after the construction of the ring of steel the City undertook an informal consultation exercise 'to ascertain the views of people and businesses who use the City', and to seek further legitimacy for their security policies.[11] Around 9000 consultation papers were delivered to businesses, residents, commuters, interest groups and neighbouring authorities between 4 August and 9 September 1993. Most responses were in favour of the City's plans to enhance security.[12] The City saw this as an endorsement of its policies, especially with regard to getting the cordon made permanent:

> We are firmly of the opinion that there is overwhelming support for making the experimental arrangements permanent, subject to a few amendments that may be required to overcome certain local problems.
>
> (Court of Common Council, 2 December 1993)

However, given the fact that this survey only attempted to ascertain the views of those working in the Square Mile, the result were not that surprising.

Fracturing the Inside Discourse

The perceived need for added security in the City after April 1993 was undeniable given the Provisional IRA's intention to continue to target the area. However, the picture painted by the Corporation of London, of the processes leading to the construction of the ring of steel, simplifies a far more complex reality. The pro-security discourse that was organised and put forward by the Corporation of London in terms of a 'collective concern' does not tell the full story of the negotiation that went on within the elite institutional networks of the Square Mile as to what the form and function of security enhancement should take. In reality there were a number of prominent parties who had different strategies as to how to cope with the terrorist threat.

In particular, the views of the City of London Police and the insurance industry differed from that of the Corporation of London. This is illustrated in Table 7.1. This table shows how different organisational frames of reference actively served to construct different 'ways of seeing' the threat of terrorism. Working through the table it can be shown how each of these groups followed a different strategy, and embarked on different relationships, both institutionally and spatially, in an attempt to achieve their objectives. This table by no means shows all the different ways in which the defensive measures employed in the City were viewed, but is intended to be reflective of the views of the groups, which were prominent in constructing the dominant inside discourse on which this book has focused.

Table 7.1 Key managerial views of terrorist risk in the City

Key Urban Manager	Corporation of London	City of London Police	Insurance Industry
Geographical strategy	Localisation for globalisation	Territorial control of space	Risk spreading
Key institutional relationships	Local boroughs and national government; The City police	Corporation of London; RUC	Corporation of London; Central Government
Key spatial relations	Neighbouring boroughs	'Collar Zone'	National focus of spreading risk
Evaluation of strategy	Cost of insurance in the city; Location and relocation	Number of terrorist incidents	Take up of coverage under Pool Re
'Way of seeing'	A 'reputational' area	A target to be defended	'Cooling' a risk hotspot

Assessing Different Spatial Strategies

Table 7.1 indicates that the strategies employed by the Corporation, the Police, and the insurers differed significantly in relation to their institutional priorities. The Corporation, for example, were concerned with maintaining and developing the well-established localised business networks which were essential for them to compete within the global economy. The key to this strategy was to create a safe and secure business location. The way in which this was assessed was through relative business migration to and from the City.[13]

The police followed a similar logic to the Corporation, highlighting the need for physical security enhancement but specifically structured on principles of 'territorial control' around the most likely target areas. As noted in Chapter 6, the police effort did not cover the entire City to the same extent. The assessment of the police strategy was simply related to the number of terrorist incidents in the City. The police and the

Corporation were in agreement that the City wanted to avoid a 'Belfast scenario' where the historical period will best be remembered for bomb damaged buildings, high levels of policing and fortified architecture. As such it was seen as important that Belfast-style security was mapped critically onto the landscape in the Square Mile given the distinctiveness of the two areas, the significance of heritage to the City, and the improved security-related technology available to the City.

By contrast the insurers were mainly concerned about redistributing the financial risk of terrorism away from the City, and hence reducing their liability in the event of another bomb in the Square Mile. The success of this strategy was judged on the willingness of businesses in different part of the UK to take out terrorism insurance coverage. The insurers' concern was on the financial and not the physical implications of a further bomb, although they were keen to offer discounts for terrorism cover if businesses took preventative measures that would either help deter possible terrorist attack or would limit potential losses in the event of further bomb.

Collaborative Relations

Table 7.1 also shows that institutional relationships were developed at a number of spatial scales by the Corporation because of the continual terrorist threat - between institutions in the City, with central Government, and with surrounding local authorities. These evolving institutional arrangements served to strengthen the City's position at a time when acts of economic terrorism were attempting to undermine it. The response of the Corporation, in this sense, could be seen to mirror wider changes in the urban governance of London that were developing at this time, focusing on notions of partnership between the private and public sectors, institutional alliances, strategic co-operation between adjoining local authorities, and vertical linkages between local and central Government. Of particular importance were the spatial relations between the City and the local boroughs as without their support the ring of steel could have been set up on a temporary basis.

The police developed relationships around anti-terrorist security with the Corporation who were their controlling local authority (for example attempting to create local CCTV associations, through schemes such as CameraWatch), and with the RUC in Belfast, whose operational experience in dealing with the threat of terrorism in Northern Ireland was influential in framing the City response. The City, like Belfast was being faced with balancing security with business normality. The Minutes of the Police Committee (27 July 1994) for example noted the positive benefits of this collaboration with the RUC:

> This Force has enjoyed a good working relationship with the Royal Ulster Constabulary for a number of years and, in particular their assistance and advice given in respect of anti-terrorist measures has been of great benefit.[14]

Of particular note were lessons from Belfast, which highlighted the threat of the displacement of risk outside the secure zone. Consequently, the police not only concentrated efforts on protecting the core of the city but also the areas outside the ring of steel – the so-called collar zone. By contrast, fundamental to the success of the terrorism insurance scheme was the relationship that developed between the insurance industry and central Government, which attempted to establish a scheme that would spread the risk at a national level. This was a difficult relationship to establish and was helped by the backing the insurers got from the Corporation of London who were keen to involve the Government in what both considered to be a national risk.

Different 'Ways of Seeing' the Risk from Terrorism

Overall, the philosophy of the Corporation showed that to protect the City's global reputation, it was forced to initiate a series of localised responses, notably security measures, to form a so-called 'ring of confidence'. The police, by contrast, viewed the City as a target which it had to defend, and adopted a strategy based on the territorial control to achieve this goal by constructing a highly visible deterrent to prospective terrorist threats. The Commissioner was also quick to point out that the fortified landscape was not just a public relations exercise to protect the reputation of the City:

> Some ill-informed people think that all we are doing is protecting those "fat cats" in the City. The reality is that if the City of London is brought down economically, perhaps never to be recovered, then all of us... will be the losers.[15]

This argument, as will be highlighted later in this chapter, was contested by some of the neighbouring local boroughs, who saw the ring of steel as dislocating the City from its less prosperous neighbours.

With regard to the insurance industry, Pool Re tried to distribute risk away from the City by dissipating it throughout the UK economy. The scheme, in short, attempted to 'cool', the risk 'hotspot' developing in the City by attempting to spread the risk nationally. This was unsuccessful, and the financial risk of further attack continued to be concentrated in the City, heightening the threat level still further. This in turn strengthened the need to adapt the ring of steel, additional risk management measures and adequate contingency planning. The relationship between insurance mechanisms and an increasingly fortified landscape was therefore important in the construction of the general pro-security discourse. Indeed, a number of respondents noted that if it were not for the ring of steel performing a City-wide risk management function it might have been difficult for occupiers in the City to get any insurance against terrorism.

Assessing the Initial Impact of the Ring of Steel

An independent property market survey in October 1993 revealed that most occupiers in the City saw the Corporation's response to terrorism as excellent and impressive (20.5%), a quick reaction (19.2%), very supportive and committed (16.4%) and effective and efficient (16.4%). Only a minority of those surveyed saw the response as slow and bureaucratic (4.8%) or chaotic (2.1%).[16] However, the results of this survey are not that surprising given the influence of business groupings in putting pressure on the Corporation to set up ring of steel in the first place. Importantly, this survey also revealed that there were a variety of different views about the operational effectiveness of the security arrangements. The results of this survey are shown in Table 7.2. This can perhaps be seen as a more realistic assessment of what the City business population as a whole felt and less of a 'knee-jerk' reaction as reflected in surveys conducted directly after the bomb, which highlighted the almost unanimous support for radical security enhancement.

Table 7.2 Opinions of 1993 security arrangements in the City

	Number	%
Doing their best	35	24.1
Will not stop the IRA	25	17.2
Welcome improvement	17	11.7
Reassuring	12	8.3
Insufficient/could do more	12	8.3
Effective/necessary	18	12.4
Haphazard/relaxed	9	6.2
Acts as deterrent	4	2.8
Reduced traffic/crime	4	2.8
Appalling	4	2.8
Detrimental to City	2	1.4
Occupational Hazard	2	1.4
Public relations Coup for IRA	1	0.7
TOTAL	145	100

Source: APR October 1993.

Table 7.2 suggests that despite the Corporation doing all they could to reduce both the terrorist threat and the fears of its institutions, the widespread support for the ring of steel should be seen against the realisation that the security arrangements are seen by a significant number as insufficient to stop the Provisional IRA if they were determined to bomb the City again (17.2%). This led a minority to criticise the police and Corporation, inferring that the current arrangements are insufficient, and shows that more could have been done (8.3%). In particular the apparently haphazard and relaxed nature of the security arrangements was criticised (6.2%). The third part of this survey also revealed that relocation by a minority of occupiers was being considered as a result of the continued terrorist threat. This survey noted that 11% in the short term and 19% of businesses in the long term were considering relocating due to fear of terrorist attack. Others indicated that they were considering a move to the outskirts of the City or to a less prominent building. However, only 3% said they were definitely moving out, but others indicated that they would have considered this option if others relocated. This view was most prevalent from those occupiers who were actually directly affected by either the 1992 or 1993 bomb.

However, the Japanese financial houses – despite calling for additional security – insisted they would be staying and did not really consider relocating from the Square Mile as the City's position in the global economy meant that they felt they had to maintain a presence. Indeed, two leading firms – Tokai Bank and the Long Term Credit Bank of Japan – that had been severely damaged in the Bishopsgate blast indicated they were not considering moving out if security was reviewed. Directors of the Tokai Bank, which had occupied four floors in the Hong Kong and Shanghai Bank outside which the 1993 bomb was parked, noted that 'London is an international business centre and doing business there is necessary…the bomb doesn't damage that image but the risk is increasing. We will have to consider how to reduce the risks for our operators'.[17]

The ring of steel and the enhanced policing procedures called for by businesses can be seen to have prevented the exodus of businesses from the Square Mile that some feared would result after the second bomb. Indeed, a year after the Bishopsgate bomb, despite some fears of relocation, it was reported in *The Times* that more financial institutions had moved into the City than had moved out.[18] The Corporation's attempts to construct a 'ring of confidence' can therefore be viewed as relatively successful. As a terrorism analyst, noted:

> One of the pros of the ring of steel is that it maintains confidence among foreign commercial companies. Nobody can deny that banking is bloody important to this country and the fact that London is the prime commercial centre between Tokyo and New York speaks for itself, and Frankfurt is increasingly going to mount a challenge, so the City is well worth protecting in this sense.

In this sense the ring of steel was seen to provide a way in which the concentration of financial services could be maintained within the spatial jurisdiction of the Square Mile, preventing an exodus to competitor locations. This was especially important as it helped maintain the physical proximity and institutional thickness of the Square Mile which was seem as vital for the success of the area.

Moving Towards Consultation

Through negotiation and collaboration between the City's key stakeholders common ground was established as to how the anti-terrorist threat should be dealt with. Initially this collaborative process was undertaken within the City's institutional networks. However, a period of statutory consultation was required in late 1993 as the initial ring of steel set up in July 1993 was only a temporary arrangement for six or possibly twelve months. Thus, through a series of local consultation and evaluation documents, the Corporation sought the views of local groups who might be affected by the proposed security measures. This was an almost continual process of evaluation and monitoring between 1993, when proposals to make the ring of steel 'permanent' were set out, and 1997, when further detailed consultation was required to extend the cordon.

The results of consultation and evaluation between 1993 and 1997 are contained in a number of important Corporation documents. Initially, *The Way Ahead - Traffic and the Environment* was presented to a series of Corporation Committees in October 1993. This was a technical and costing assessment of the experimental arrangements, with the proposed aim of making the scheme permanent. It also highlighted the links this scheme had with the Corporation's UDP, and in particular the proposed Key to the Future scheme, which was seen to complement the access restrictions introduced to reduce terrorist risk. This 1993 report also set out the consultation framework that would need to be followed for the ring of steel to be made permanent.

Further results from this initial consultation came in the form of *The Way Ahead - Traffic and the Environment - Results of Statutory Consultation and Public Notice in early 1994*.[19] This report sought to advise the Corporation on the results of the public notice and the statutory consultation with the neighbouring boroughs, emergency services, representatives of road users, and other members of the public. It also updated the ongoing assessment process with regard to traffic and environmental issues, advising slight modifications to the temporary scheme, but not unsurprisingly, given that the Corporation undertook it, the report fully endorsed that the ring of steel be made permanent.

A year later, a further report presented to a number of Corporation Committees in March 1995 entitled *Traffic and Environmental Area Suggested Western Extension* highlighted the ways in which the present ring of steel could increase in size to encompass more of the Square Mile. The proposals were twofold. First, a westward extension to the ring of steel *per se* to encompass the areas of St Paul's, Smithfield and the Barbican. Alternatively, the other suggestion looked at the possibility of creating a number of localised traffic and environmental improvements schemes in different areas of the City, which could, at some point, be linked together into one major zone. After the Docklands bomb in 1996 the former of these plans was highlighted as the preferred solution. A further report was then commissioned called *Suggested Western Extension of Traffic and Environmental Zone* where the City Engineer asked the Corporation for permission to extend the ring of steel based on a request from the City of London Police (Policy and Resources, 17 October 1996). This was eventually implemented leading to a subsequent evaluation report having to be written called *Traffic and Environmental Zone - Western Extension - Evaluation Report* which was a technical assessment by the City Engineer on extending the ring of steel. This was presented to a number of Corporation Committees in May/June 1997 and sought to make the extension permanent under the guise of local environmental improvements.

As a result of this general consultation process two findings frequently emerged which the Corporation used to support the introduction of security measures, neither of which were directly related to security enhancement - the benefits of the ring of steel for road users and pedestrians, and environmental improvements brought about inadvertently by security enhancement.

Traffic Benefits

The maintenance and enhancement of the transport infrastructure in the City, as noted earlier, was a key concern of the City, given the expansion of its office stock since the 1980s. Therefore the transport-related benefits of the ring of steel were of significant importance for the Corporation. Initial computer records of transport patterns in the City during 1993 showed there was a significant reduction in traffic in the central areas of the City as a result of the changes to traffic routes introduced by the security cordon. However, according to Corporation records, the level of traffic in the City did not change, indicating that the displacement of traffic had been contained within the Square Mile. The changing transport routes also meant that 18% of the bus routes into the City were affected, although most of the companies contacted reported no loss in trade. The same was true of coach firms and taxi drivers, although isolated complaints were received. Furthermore, consultation between the Corporation and British Rail, London Underground and the emergency services reported no initial problems with the new security scheme. Consultation with these bodies continued for a few months so that the implications of the construction of the ring of steel could be looked at over a longer time-scale.

The 1994 report to the Planning and Transportation Committee qualified the previously announced trends. It showed a 25% reduction of vehicles entering the central areas of the City with an 18% reduction in journey times. Consultation with occupiers of the City at this time also indicated that most did not perceive that there had been a significant reduction, or worsening, in traffic noise or pollution. Furthermore, a noticeable reduction in serious road accidents occurred.

With regard to public transport, in 1994 the Corporation found no negative effects and in some cases, significant advantages accrued. For example, it was reported that some bus routes benefited from up to 70% reductions in journey time. However, the results of the public consultation exercise showed that taxi drivers were increasingly concerned about access to certain areas of the City, which they felt hindered their business. Furthermore, one major objection did arise with regard to the displacement of traffic from the City into areas under the jurisdiction of Tower Hamlets borough council. This will be highlighted later in the chapter.

The 1995 City Engineer's report detailing proposals for an extension of the ring of steel noted that computer modelling indicated that the extended area in the west of the City should see a further 10% reduction in traffic with no effect on the original area. Police statistics also showed that road traffic accidents were continuing to decrease. The 1996 Engineer's report went further, indicating that no traffic should be displaced onto roads of neighbouring authorities. The evaluation report of the implementation of the extension qualified these predictions with the total traffic passing through the City staying the same. Prior to this the minutes of the Policy and Resources meeting (April 1996) noted that:

> It is considered that, in essence, the impact will be analogous to that created by the introduction of the 1993 Traffic and Environmental Zone. Some unavoidable inconvenience to vehicles requiring access to certain streets is inevitable. However, a vital aspect is the ability to contain the displaced traffic on the City's Secondary and Local

> Road Network, with a neutral effect on roads in neighbouring boroughs with whom close consultation will be required.

Overall, the road users who put forward opinions to the Corporation saw the ring of steel as a primarily positive traffic management feature although most were well aware of its anti-terrorist applications.

Environmental Benefits

As previously noted the ring of steel was, from its inception, seen as beneficial to the City's environment. This was in line with the Corporation's environmental policy 'Key to the Future' which aimed to cut traffic congestion and pollution in the City. The 1993 consultation report showed that noise and atmospheric pollution were reduced in the central City as traffic volumes were reduced and many side roads closed to traffic. For example, a 12% reduction in nitrogen dioxide was reported in the central City in the first two months that the ring of steel was in operation. This was further confirmed by the 1994 report, which showed that overall pollution levels within the cordon had decreased by around 15%, and that pollution in the City as a whole had decreased slightly. This report again highlighted other benefits to the environment, such as reduced noise pollution and decreases in the soiling of buildings (important given the number of listed buildings in the City), due mainly to an 18.1% decrease in particulate pollutants such as lead. This report, however, did qualify these findings by pointing out that traffic (and noise pollution) in other parts of the City might have increased as a result of the cordon. Additionally, this report noted that, if the cordon were made permanent, continual environmental improvements would probably accrue.[20]

In the 1995 report by the City Engineer on a possible extension to the ring of steel, the same arguments as noted above were used by the Corporation - namely that the environment would benefit from a reduction in traffic flow and that this would provide an opportunity to enhance the City's street scene. As it concluded:

> A scheme can be implemented that will extend the traffic and environmental benefits experienced in the core to a larger area of the City of London encompassing the environmentally sensitive areas of St Paul's, Smithfield and Barbican.

The City Engineer, downplaying the proposed anti-terrorist benefits, further argued that: 'the scheme is operating efficiently and substantial environmental benefits have already resulted, with powerful enhancement of the City scene.' He believed that 'there was a strong case to build on our past experience and extend the environmental benefits further to the west by developing further complementary schemes currently in various stages of development'.

When formal plans were laid out for extending the ring of steel in the 1996 report after the Docklands bomb, environmental and transport justifications were officially given, not the continuing terrorist threat, with security enhancement noted only as a beneficial by-product. This as argued previously was an attempt to shift the emphasis away from the enhancement of security. The 1996 report also noted that the extension to the ring of steel would divert or discourage a further 10,000 vehicles from entering the City centre each day.

The 1997 report showed that the experimental extension to the ring of steel had had major environmental benefits in the new secure zones in terms of noise and pollution reductions. The 1997 Corporation of London publicity document on what was officially called the Experimental Western Extension of the Traffic and Environment Zone was distributed to all businesses within the Square Mile as well as in neighbouring areas, and summarised the Corporation's view that the ring of steel was advantageous without making any reference to anti-terrorism strategy:

> The original scheme was highly successful in improving conditions within the City's central area. People have praised the reduction of traffic and better quality of the environment. Since the arrangements were made permanent...additional planting, seating and new paved areas have been introduced, making the City an even better place to live and work.

Contesting the Inside Discourse – Objections from Outside the City

Whereas the ring of steel was undoubtedly of benefit to improving the internal environment of the City and in reducing the fear of further terrorist attack (even if this benefit was downplayed), such arguments do not account for the views of those excluded from the initial consultation processes that highlighted the potentially negative aspects of the cordon. The views presented in Table 7.3 are mainly from official bodies who had direct links with the Corporation of London. It is therefore not surprising that they supported the Corporation's plans with regard to the ring of steel.

As well as legitimating the ring of steel and other security enhancing measures, the consultation and evaluation process also allowed many groups and organisations outside the Corporation of London's institutional nexus to comment on what the security measures employed meant for them, as formal consultation was legally required after the ring of steel had been in place for six months. Thus in December 1993 the Corporation set out to explain the Corporation's proposals and asked recipients for any concerns they might have (Planning and Transportation Committee, March 1st 1994). Public notices were placed in the *Evening Standard* and *London Gazette* (21 December) and consultation letters sent to neighbouring boroughs, the Metropolitan Police, the Emergency services, public transport operators, and the London Docklands Development Corporation.

The results of the Public Notice were still, on the whole, generally favourable to the ideas for security enhancement put forward by the City. The impression the Corporation were trying to create – of unanimous support for its scheme – was further enhanced by the views of a host of other predominantly City-based organisations and suggest almost unequivocal support for the security proposals, further reinforcing the Corporation's claims that the majority of people were in favour of their proposals.

For the first time the Public Notice, allowed the neighbouring local boroughs and other interested parties, to *officially* have a say. This process yielded views, which, in some cases, were in direct opposition to those expressed by the Corporation. Furthermore, objections to the security enhancement were also raised, contesting the view of the inside discourse that the security scheme would be mutually beneficial and have no knock on effects in areas outside its boundary. The 'outside discourses' that emerged questioned the assumptions of the Corporation about how the anti-terrorist effort should be dealt with. Opposition was centred on views of neighbouring local authorities (in particular Tower Hamlets) and civil liberties implications. There were also those in the wider community that questioned the operational effectiveness of the ring of steel. The remainder of this chapter will detail these different aspects of the outside discourse.

Table 7.3 Additional responses to the public notice (March 1994)

Grouping or organisation	Response to the Public Notice
London Docklands Development Corporation	Concerns over traffic between the City and Docklands areas
Metropolitan Police	No objections
London Fire Brigade	No objections
London Ambulance Service	No objections
London Transport Planning	No objections
London Buses	No objections
London Regional Passengers Committee	No objections
British Rail	No objections
Bus and Coach Council	No correspondence
Cyclists' touring club	Slight problems raised
The City of London Environment and Amenity Trust	Objection
Public objections to cycling access	Two objections received
Licensed Taxi Drivers Association	Objection
The (Taxi) Owner Driver Society	Objection
Organisations representing people with disabilities (x 5)	No objections
The City Retail Traders	Welcomes the measures
London Chamber of Commerce	Supports the measures
The Freight Transport Association	No objection
Utility companies (x3)	No objections
The Royal Mail	No objections
The Barbican Association	No objections

The Initial Views of the Neighbouring Boroughs

After the ring of steel was established, the neighbouring boroughs expressed their grievances, noting that they had, in their opinion, not been consulted properly on what the City was planning to do, and that the security cordon could have negative impacts on surrounding areas in terms of increased traffic and the possible displacement of terrorist risk. Furthermore, there were concerns that the scheme had little co-ordination with the policies (especially transport) of the neighbouring areas.

Objections immediately came (July 1993) from some neighbouring boroughs. Tower Hamlets wrote to the Government and the Corporation over the complete lack of consultation and Islington highlighted that it had only heard of the cordon by fax the night before it was implemented. Southwark also wrote to the Corporation regarding the inadequate consultation and pledged support for Tower Hamlets' objection to the Government. The London Docklands Development Corporation further objected to the scheme as they felt it made travel between the City and the Isle of Dogs increasingly difficult, as traffic to the east of the City (in Tower Hamlets) would increase. As a security officer in the Docklands, indicated, the neighbouring boroughs were concerned about how the ring of steel would affect them:

> I recall attending a meeting over in Canary Wharf after it [the ring of steel] went up to discuss the actions taken by the City in relation to the surrounding boroughs. All the surrounding boroughs could do was sit there and criticise them for putting it out without giving any thought to the knock on effects of traffic.

Tower Hamlets and the other boroughs were not against the idea of a ring of steel *per se* but they were concerned about how it could adversely affect their area in particular, and movement policies for central London as a whole. As a spokesperson for Tower Hamlets indicated:

> We are not saying they need to rethink the seal, but the City of London is not completely separate. The Corporation can not do what it likes without thinking about the impact its moves may have on the neighbouring boroughs.[21]

The objections of many of the boroughs were articulated through the Labour controlled Association of London Authorities (ALA).

Others saw the actions of the Corporation as paternalistic and an attempt to spatially imprint their ideas onto the landscape of central London, especially in relation to transport policies. When the ring of steel was implemented, Andrew Pharaoh, the director of 'Movement for London', the London arm of the British Roads Federation, indicated that severe disruption to normal traffic flow would occur: 'It is going to cause massive problems around the borders of the zone, especially if other boroughs introduced their own schemes.'[22] Similarly, the Islington council leader indicated that the implementation of the ring of steel had not been thought through properly, noting: 'there is a clear need to protect people from terrorism but this plan is completely ill-conceived.'[23] Furthermore, the ALA indicated that consultation is the key to a successful scheme: 'If one small part of

London takes action without co-ordinating the plan with other boroughs it could lead to chaos.'

Although there was obviously strong resentment about adverse traffic effects and the lack of official consultation between neighbouring boroughs and the Corporation of London before the ring of steel was implemented, such criticism must also be seen within the wider context of a fragmented London-wide transport policy. During the early 1990s proposals for changes in London transport management were causing borough councils and local pressure groups to oppose any new traffic measures that would introduce more traffic flow to their areas. Traffic management for London at the time of the introduction of the ring of steel was shared between the boroughs and the Department of Transport, which gave 'tangible political expression to the conflict between neighbourhood and wider issues' (Travers *et al*, 1991, p.102). In order to alleviate this tension, the Chartered Institute of Transport noted in June 1991 that:

> Transportation policies pursued in one borough can directly affect adjacent boroughs and there is a general agreement that it is essential for transport policy to be undertaken in a co-ordinated framework.

The construction of the ring of steel by the Corporation was certainly against this ethos, even though it did not officially have to consult with the boroughs for six months, as it was presented as a 'temporary traffic management scheme'.

In October 1993, the City Engineer, in support of the Corporation's proposals to make the ring of steel permanent, indicated that 'as yet, no detrimental traffic effects have been identified in surrounding boroughs as a result of the current experimental traffic arrangements'. However, in this report there was also an admission that the surrounding local authorities felt it was too soon to judge this as the scheme had only been operational during the summer holidays and did not adequately reflect normal City of London traffic flows and work patterns. The report continued by noting the concerns of the neighbouring boroughs and their wish to be fully consulted:

> Some authorities have expressed a view that traffic flows have changed in their area, but no one is yet in a position to determine whether this was due to the City scheme... [and] some have also expressed the wish that the Corporation of London should attempt to consult more individuals in the surrounding local authority areas.

Subsequently the 1994 report, which gave the result of the statutory consultation exercise, indicated that experiments carried out in October 1993 showed no clear patterns linked to the ring of steel could be proved. However, it was noted that certain roads in Tower Hamlets had received noticeable increases (up to 8%) on certain roads.[24] This was not an unexpected result as Tower Hamlets had been indicating that this had been occurring for some time. This, as will be discussed in the next section, formed the basis of significant objections to the ring of steel when the statutory consultation period began in early 1994.

Official Consultation with the Boroughs

Given the limitations of the Road Traffic Act, the City based its legal case for renewal of the ring of steel (in 1994) on transport and environmental grounds. Therefore, between the approval of such measures at the Court of Common Council on 2 December 1993 and publication on 21 December, senior representatives from the City Engineer's department visited all seven neighbouring boroughs and the London Docklands Development Corporation to explain the proposals and discuss concerns they might have. On 21 December consultation letters were sent out to the seven neighbouring authorities. This provided a twenty-one day objection period. This three week period was however extended to eight weeks so that the local authorities could properly consult with their members. The responses of the authorities are shown in Table 7.4.

Table 7.4 Views of the neighbouring authorities on the ring of steel

Neighbouring Authority	Consultation Comments
City of Westminster (west of City)	No objection to making the scheme permanent.
Islington (north of City)	Holding objection with concern expressed about 'geographical' areas that would be expanded upon subsequently.
Camden (north-west of City)	No objections with the proviso that if any disbenefit with Camden occurs ameliorative measures will be sought from the Corporation.
Hackney (north-east of City)	No objections but would like to see better cycle access. Slight concern also expressed about the effect of Police activities on buses.
Tower Hamlets (east of City)	Objection on traffic management grounds.
Southwark (south of City)	No objections unless London Transport or the emergency services lodge complaints.
Lambeth (south-west of City)	No objection to making the scheme permanent.

Table 7.4 indicates that despite initial criticism of the scheme (in large part to do with the fact they were not consulted), the majority of the neighbouring authorities were happy with the ring of steel becoming permanent as long as it did not impact upon their area.

The holding objection that Islington had was immediately dropped. This objection concerned the possible expansion of the scheme to encompass the Broadgate centre, which was to become part of the City in 1994. Whilst the original ring of steel did not impact upon traffic flows in Islington, the borough council were concerned that detrimental traffic flow patterns might occur if the ring of steel was expanded northwards to cover Broadgate.

With Islington dropping their objection this left Tower Hamlets as the sole objector from the neighbouring boroughs. This was a potentially serious matter for the City, as if this objection were not withdrawn a public enquiry would have been forced at great cost. For example, the Policy and Resources meeting on 10 February 1994 noted that the Corporation was concerned about the possibility of a public inquiry and had organised for MORI to carry out research on the public's view of the Corporation's proposals before any possible inquiry, so that it could highlight what it felt would be unanimous support for its ideas.

On 4 March 1994 the Corporation of London asked the Secretary of State for his determination on Tower Hamlets' objection. At this time a number of meetings between the City Engineer and representatives of Tower Hamlets and the Minister for Transport in London took place. These meetings tried to resolve the specific objections Tower Hamlets had, related to what they saw as substantial increases in traffic flow on their roads, to the east of the City, caused by recent modifications made to the road network in their area:

> The principle concern of Tower Hamlets is the current traffic situation immediately East of the City boundary which has significantly affected the opening of the Limehouse Link on 17 May 1993. This new radial link road has resulted in substantial increases in traffic arriving at the Tower Hamlets/City boundary and discharging into a road network not capable of adequately coping with it. The introduction of the experimental traffic and environment scheme on 3 July 1993 by Mr Commissioner had been perceived as a compounding problem by [the] London Borough of Tower Hamlets.[25]

After great pressure had been put on Tower Hamlets, a 'Memorandum of Understanding' was reached with a series of remedial measures planned which would attempt to ensure better traffic control on the mutual boundary. This was a three-way agreement between the Corporation, the Government and Tower Hamlets:

> On the basis of this 'Memorandum of Understanding' and a further letter of support from the Minister for Transport in London, the London Borough of Tower Hamlets has withdrawn its objection to the proposed permanent Traffic Orders for the City traffic and environment scheme.

The objection was therefore withdrawn allowing the ring of steel to be made permanent on 2 July 1994, a year after its initial implementation. Michael Cassidy indicated that Tower Hamlets were not easy for them to liaise with during this process:

> We needed cross-borough co-operation and Tower Hamlets have not given it priority over the last three years. They raised strong objections at the implementation stage and were the last to withdraw their objections...Tower Hamlets dropped objections to the scheme at

the last possible moment and a public enquiry was avoided. The City in return offered 'various assistance' to Tower Hamlets.

The next time major changes to the ring of steel were suggested by the Corporation – to extend the ring of steel after the Docklands bomb – again, only internal consultation was required. The extension was ratified in December 1996 and introduced on 27 January 1997 for an experimental period of twelve months. According to the Corporation:

> During this time [the experimental period] full consultation will take place to seek the views of those people who live and work or have businesses in the extended area. At the end of the consultation period all views received will be collated and carefully thought about by the Corporation when it contemplates making the extension permanent.[26]

This makes no mention of wider consultation with the surrounding boroughs, and as with the 1993 cordon, the planning of the extension was criticised by the neighbouring Labour councils who accused the Corporation of again failing to consult with them adequately. Islington were particularly aggrieved as they had foreseen this extension being implemented when they initially put in a holding objection to the standing order in early 1994. For example, a spokeswoman for Camden and Islington Councils noted that:

> The City is using police powers to avoid going through the normal planning procedure. It is distributing its leaflets now telling people it has already decided on this.[27]

Corporation of London Minutes confirm that it was only on the day of implementation that leaflets were distributed in Islington and Camden. However, they also indicated that a meeting had taken place between them and Islington members after the extension was put in to discuss the traffic effects, which were shown to be minimal.[28]

Indeed, in late 1997 there was talk of a further extension to the ring of steel to cover more of the area around the Broadgate centre. There were feelings expressed by the Corporation of London and the surroundings boroughs of Hackney, Islington and Tower Hamlets that the fringes of the City were vulnerable to attack. It was reported in the Daily Telegraph that the Corporation was considering 'shoring up the exposed flanks' by altering the traffic flow in these areas in early 1998.[29] The merits of this proposal can perhaps be shown by the attempt of an Irish Republican terrorist group to set off a series of firebombs in Central London in July 1998. One of the bombs was reportedly found at Chancery Lane underground station on the western border of the City. This extension was also planned with the idea of reducing traffic volumes in this area of London. For example a report in early 1998 indicated that total car volumes had decreased by 25% in the City since the ring of steel was introduced in 1993.[30]

In summary, the consultation process with the surrounding local authorities saw the establishment of some institutional links between the City and its neighbours. Similar institutional links were also being made at this time in other policy areas, as there was a considerable emphasis on voluntary co-operation between London boroughs for London-wide benefits. However, there was also a feeling that the needs of the boroughs were being marginalized, as emphasis was placed on London's role as a global city and in particular the City's place within this. The implementation of the ring of steel exemplified the powerful nature of the City, and the almost uncritical acceptance of its 'inside' discourse, which was supported by Central Government but criticised by the neighbouring boroughs who were less concerned with 'world city' agendas than with local priorities.

Civil Liberties Implications

Despite the ring of steel improving business confidence in the City, some individuals and groupings, outside of the formal consultation process complained to the City Police and Corporation about its side effects. As soon as the ring of steel was erected in July 1993 there were complaints about the scheme being an infringement on civil liberties, making everyone a potential suspect, and an encroachment on the public realm with each layer of security seen to be imposing further restrictions on the public's freedom. These complaints centred on three key concerns that are symptomatic of general complaints made against crime prevention measures: first, restrictions on movement and access to public spaces; second, the freedom from suspicion and the ability of police to stop and search; and third, the omnipresent surveillance coverage by CCTV.

A number of organisations were concerned that access restrictions to areas of the City would affect them economically. For example, the reduction in transport routes into the City was seen as detrimental to public transport operators. Table 7.3, for example, showed that there was some concern, most notably from taxi drivers that access to the City was going to be restricted.

Furthermore, civil liberty groups were also critical of changes in the law, which allowed people to be randomly stopped and searched when entering the ring of steel when they had not committed any actual criminal offence. This reverses a fundamental principle of English law, that a stop and search can only be made on suspicion. John Wadham, Director of Liberty, noted in 1993 that 'we believe the balance as it is now is about right. We understand why the police want more powers but do not think their powers should be increased'.[31] In particular he believed that the new security arrangements would lead to harassment of the Irish Community in London. Wadham further noted a year later in a letter to *The Times* in April 1994 that he believed the police were using the security cordon for measures not associated with terrorism:

> The statistical evidence for the ring of steel...shows that police have stopped a disproportionate number of black people, despite the fact that to date no black people are believed to have been involved in the IRA activities in Britain.

There were also been those who have complained about the imposition of surveillance cameras:

> There are always complaints about access restriction but I have also heard complaints about the invasion of privacy some people feel when they go though the cordon due to the CCTV and knowing they are being watched.[32]
>
> (Senior City Risk Manager)

Civil liberty complaints were, on the whole, uncommon but provided a viewpoint that the Corporation did at least consider, even though they were marginalised and had little effect on the way in which the Corporation and the City Police developed the ring of steel. Civil liberty considerations, like those of the borough councils, were of minimal importance to the Corporation in their efforts to construct a 'ring of confidence' for business occupiers and to safeguard the City's reputation.

Questioning the Effectiveness and Usefulness of the Security Cordon

The extent to which the ring of steel genuinely provided a safe and secure business location, was also contested by a number of individuals and organisations, who questioned both its operational effectiveness and the arguments initially made by the Corporation that the ring of steel was creating a 'secure zone' within the City.

Many occupiers saw the ring of steel as an almost total guarantee of security. Its merits were further enhanced through the consultation process and subsequent Corporation reports which attempted to legitimate this by pointing out the positive benefits of the scheme in terms of reducing crime and terrorism, as well as environmental and traffic benefits.

However, the Provisional IRA were certainly convinced that attacking the City was still a worthwhile tactic and that the ring of steel was ineffective. A letter to City business (dated 8 July 1993) that they sent after the ring of steel was announced indicated:

> No one should be misled into understanding the seriousness of the Provisional IRA's intention to mount future planned attacks in the political and financial heart of the British state. Furthermore, no one should allow the futile announcement of a City of London security zone...to lull them into a false sense of security.

Additionally, other commentators believed that the security operation provided an incentive for the Provisional IRA to demonstrate that the cordon could be breached. This view suggested that far from being the answer to the City's terrorist problem, the ring of steel increased the risk of further terrorist attack as media exposure in the event of a successful strike would be of propaganda value to the Provisional IRA. For example Dr Conor Cruise O'Brien, a leading writer on Irish affairs, writing in *The Times* newspaper, indicated that in his opinion:

> The ring of steel increases the risk to the City in two ways. It increases the incentive to the IRA to strike, because of the propaganda value to be derived from penetrating that loudly trumpeted ring. The other way in which the charade increases the risk to the city is that it diminishes manpower available to counter the IRA threat. Fixed roadblocks need a lot of trained manpower.[33]

The Commissioner of Police indicated that at this time he had been getting criticised for throwing a gauntlet down to the terrorists. He defended his actions by noting that: 'given the history of what they have done to the City, one has to wonder what more of a challenge they needed' (Kelly 1994a). Others also questioned whether the ring of steel provided an adequate defence against the Provisional IRA:

> The ring of steel is not even a serious attempt to stop the bombings. The security measures currently in place are insufficient to present sufficient deterrent to make someone think 'that's too dangerous, I'm going somewhere else'.
>
> (Senior City Risk Manager)

He continued by ironically indicating that he felt that offices located outside the security ring are in fact at an advantage as they avoid much of the traffic congestion the cordon causes – 'I think people were quite relieved they were outside the ring of steel as they don't have the hassle of having to get through it to get to work'.

Those who cast doubt on the effectiveness of the ring of steel served to contest the meaning that the City authorities had attempted to place upon their security operation. Furthermore, the dominant pro-security inside discourse was challenged as the perceived terrorist threat declined. After the Provisional IRA called a ceasefire in August 1994, some sections of the media were calling for an immediate scaling down of the City of London's ring of steel, indicating that in their opinion the threat had gone away overnight. Indeed, the City Police reported that they found it difficult to maintain a 'culture of security' in the area due to the apathy of the community about the risks they faced. Subsequently the Corporation came under pressure from sections of the business community to down grade security. This pressure intensified as the ceasefire progressed, and reached a peak towards the end of 1995 after the ceasefire had been in place for over a year:

> It came to a point just before last Christmas (1995) where there was a very real effort on behalf of a number of organisations and prominent individuals to actually get it back to where it was before. I mean we gradually sort of eased off but our profile had been reduced although all our measures were in the back cupboard waiting to be put in place again.
>
> (Senior City of London Police Officer)

He continued:

> And it came to a very sort of final negotiation to the position to be readopted, in the early part of February, and then the Docklands bomb occurred and then everyone was saying 'thank Christ we didn't do it'.

This highlighted that there were sections of the business community who at this time wanted the ring of steel disbanded, with the City returned to pre-St. Mary Axe days where there were no overt signs of anti-terrorist security on the City streets. However, after the Docklands bomb in early 1996, the risk-profile of the City was again high and instead of backing a reduction of security, the business community, as a whole, was keen to get the cordon extended westwards.

Conclusion

The threat of terrorism facing the City in the 1990s increasingly sought to focus attention on how the Corporation was adapting to modern conditions, relating to its role in helping to promote London as a worldwide city, as well as maintaining its influence within the global economy. The ring of steel helped the City to create a secure platform upon which it could continue to develop and adapt its role as the financial heart of a 'world city'.

Despite its tradition, the City during the 1980s and 1990s showed itself to be a thoroughly modern organisation in the way it adapted its agendas. This was especially true after the abolition of the GLC in 1986. As Hebbert (1998, pp.120-121) noted after 1986 the Corporation... 'took over most of the work of representing London overseas and receiving important visitors to the capital... Taking its seat on various new joint committees of the London boroughs, it used its current status – venerable, immensely rich, a political eunuch – to broker between the ideological blocs of boroughs controlled by Labour and the Conservatives...Happenings that would have seemed wildly incongruous ten years beforehand became commonplace'.[34] The development of the ring of steel can perhaps be seen as a physical manifestation of these sentiments and the evolving role of the City in London-wide agendas at this time.

As has been noted, after the Bishopsgate bomb the need for the construction of radical security measures was articulated by a strong pro-security discourse as a variety of key institutional actors got together to mobilise support for security changes based on 'collective concerns' for terrorist risk. In particular, the Lord Mayor and the chairman of Policy and Resources, Michael Cassidy, became key voices in the City's response to terrorism (just as they were in the promotion of London as a whole), although the powerful financial institutions were still very influential in shaping Corporation policy and practices. Subsequently, the pro-security (inside) discourse was legitimated, at least in the eyes of the Corporation, through the consultation process. There was also a significant degree of negotiation and contestation from the disempowered 'outside discourses' over the Corporation's approach.

The various meanings attached to the ring of steel meant that it was viewed in a variety of ways by different organisations and groupings: a Belfast-style anti-terrorist cordon (by the media); a territorial policing strategy (City Police); an effective traffic management measure and beneficial environmental approach (the Corporation's official line); a 'transferer' of traffic to adjacent areas (neighbouring boroughs); a public relations exercise to stop businesses leaving (critics of the cordon); an imposition of personal freedoms (civil liberty organisations); an insurance necessity (insurance industry); a beneficial measure helping promote London as a 'world city' (Central Government); an exporter of the risk of terrorism, or simply as an effective or ineffective measure against terrorism.

The construction of the ring of steel was planned and executed along the lines proposed by the Corporation of London who either used other 'readings' of the landscape to legitimate the security cordon (environmental improvements, traffic management and crime reduction), or simply ignored them (displacement of traffic and civil liberty complaints). In short, the City was able to construct a ring of steel because of the powerful inside discourse that emerged based on heterogeneous agency in support of radical security proposals. Attempts to construct rings of steel in other areas of the country at this time (most notably Manchester) shed light on the key processes and powerful operation mores which developed in the City, showing the extent to which the ring of steel was based on wealth, power and a strong institutional thickness with a shared discourse of action.

It can be argued that the risk of terrorism and the subsequent response in the form of a ring of steel actually served to enhance the City's institutional thickness through the construction of new networks of relations at a variety of spatial levels. At a local level, the Corporation increasingly liaised with businesses for security purposes and encouraged businesses to co-ordinate their efforts for mutual benefit. New City-wide committees were set up to feed in to the Corporation's response to the terrorist threat. At a wider geographical scale institutional relationships were also established with neighbouring boroughs as well as Central Government, which entering the 1990s would have been unthinkable given the detached role the City had traditionally played in London affairs.

However, it was the wider driving forces of the global economy that ultimately framed the response of the Corporation to terrorism, which was physically expressed in the ring of steel. The Corporation was well aware that it needed to reduce the perceived vulnerability of the City to acts of economic terrorism in order to enhance its position as a global trading centre, but also importantly, London's position as a 'world city'. It did this by attempting to balance security with the effective functioning of business by providing a 'ring of confidence' for national and international businesses.

Indeed it was not long after the ring of steel was constructed that the landscape changes that occurred in the City were highlighted as improving the quality of life in place promotional material for London in which the City was portrayed as a safe area in which to conduct business:

> The City of London 'Square Mile' has benefited particularly from a cut in crime following the reduction in entry points...following a bomb in 1993...The unplanned result of the initiative has been a cut in office burglaries in the Square Mile and a corresponding increase in business confidence.[35]

However, whilst the risk of terrorism forced the City to look outside the borders of the Square Mile for institutional liaisons, the motives of the Corporation as a local authority can still be viewed as paternalistic, protecting itself first and foremost. Its inclusive looking agenda, whilst helping it connect to global markets, also led to it attempting to disconnect itself in the physical sense, increasing its boundedness and physical separation from the rest of London at the same time as London-wide partnerships were increasingly being forged.

In its business dealings the City has always tried to exist on a plane above the physical city, leaving the City of London in a disengaged relationship with its own neighbouring areas. Therefore it was perhaps ironic that the risk of terrorism that afflicted the City during the early to mid 1990s, far from weakening the City's trading position, served to bring together firms, institutions and local and central Government, all anxious to defend the Square Mile even though the ring of steel served to fragment the physical landscape of central London.

Entering the new millennium the City has continued to enact change to remain competitive as a world city. It continues to be central to London affairs, especially around economic competitiveness agendas and closely linked to the new Governance structures for London such as a creation of a Major and Assembly for London - collectively referred to as the Greater London Authority (GLA) (Tomaney, 2001). For example the Corporation of London's Chair of Policy and Resources has become the Mayor's informal advisor on business.

The City's changing position has also been expressed physically, demographically and culturally. The City continues to adopt a policy of expansion both upwards, through a 'tall buildings policy' and outwards through the development of City Fringe sites. As noted in Chapter 5, during the early 1990s what the financial Corporations required was large 'groundscaper' buildings which were low buildings with huge floors to accommodate large financial dealing rooms. Entering the new millennium the craze was for high profile skyscrapers. Perhaps the best example of this is the new Swiss Re building designed by Norman Foster for the site of the 1992 St. Mary Axe bomb (see Figure 7.1).

The tall building policy in the City and surrounding areas was pushed hard by the GLA and the City for mutual benefit. The City required new buildings to stave off a fresh challenge for man ever expanding London Docklands whist the GLA could insist that the Corporation financed the buildings of social housing stock as a 'sweetener' for being given planning permission (McNeill, 2002). As such the Corporation of London is increasingly entering into partnership deals with surrounding boroughs in a way unheard of ten years ago (Sudjic, 2001a). For example the Corporation of London's Economic Development Action plan for 2002/3 noted that:

> In pursuing economic development... the Corporation [need] to demonstrate its overall relevance in relation to economic, social land environmental developments in London – that is London as a whole: the surrounding boroughs; and in the Square Mile itself.

After 9/11 questions were inevitably asked about the future of tall buildings and trophy skyscrapers that can not be constructed to withstand a direct hit from a plane or that take too long to evacuate (Archibald *et al*, 2002; Hall, 2001; Williams, 2001). However, it appears that in the City of London 9/11 has not impacted upon long term office development policy because as Bowers (2002) notes, 'there is no defence against a September 11-style terrorist strike short of living in bunkers'.

Figure 7.1 The Foster-designed building on the site of the St. Mary Axe bomb

The Square Mile also continues to increase its residential population as well as the international presence of overseas firms and as such is widening its structures of governance and hence destroying the last remnants of the 'club atmosphere' for which the Square Mile was renowned (Kynaston, 2001). Others however believe change has still not gone far enough and that the governance structures in place are still undemocratic and inappropriate within wider London governance structures (Power, 2001).

The Automated Number Plate Recording (ANPR) technology utilised in CCTV which developed throughout the City's attempts to deter Provisional IRA terrorists, is also being 'rolled out' across central London for use in traffic 'congestion charging' which started in February 2003.[36]

All number plate images are captured when entering the zone by ANPR technology and automatically matched against database of those registered to pay or have exemption. Other cameras monitor the general flow of traffic throughout the zone. There are also mobile camera patrols operating throughout the zone. The zone is surrounded by an inner ring road which is free of charge. The specifications of the scheme are detailed below in Table 7.5.

Table 7.5 Characteristics of the Congestion Charge scheme

Key Data of Scheme	System Specification
Sponsor	Transport for London
Budget	£200 million
Expected annual revenue	£130 to £150 million
Annual running cost	£50 million
Expected profit	Approx. £80 million
Area of charging zone	8 square miles (21km²)
Current traffic volume entering zone (peak)	40,000 vehicles per hour
Entry and exit points	174
Number of entrances per day	400,000 (250,000 separate vehicles)
Camera positions	230 (each with up to eight cameras)
Total camera numbers	450
Camera height	8ft
Camera success rate	90% accuracy reading number plates
Charge times	7am to 6.30pm Monday to Friday
Cost of daily pass	£5
Cost of weekly pass	£25
Cost of monthly pass	£110
Cost of annual pass	£1,250

The City itself is on the eastern border of the zone (Figure 7.2). In essence Inner London will be circled by digital cameras creating a dedicated 'surveillance ring' affording not just the City Police but also the Metropolitan Police vast surveillance gathering capabilities for tracking the movement of traffic and people, and by inference highlighting potential terrorist threats.

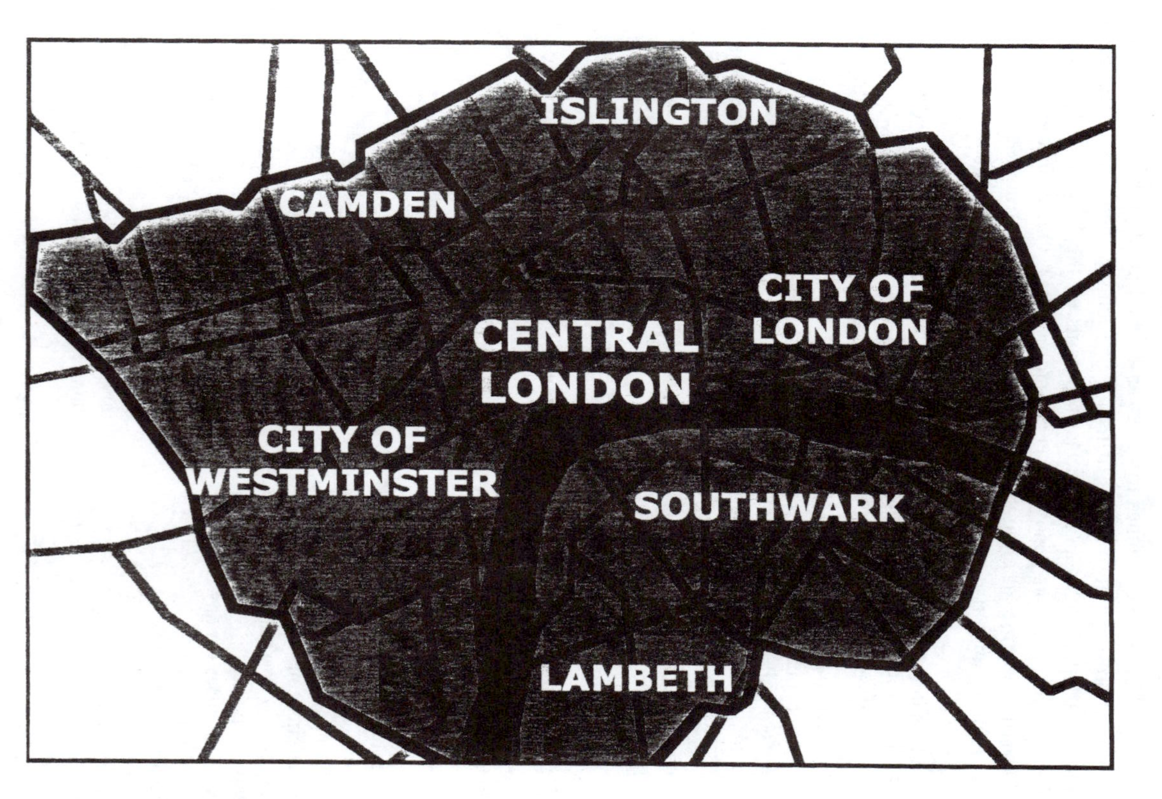

Figure 7.2 The City and the new central London congestion charge zone

Figure 7.3 The border of the congestion charging zone on the eastern side of the City of London

Figure 7.4 Entry point security cameras in the City of London

Not surprisingly such an anti-terrorist function for the new congestion zone has been largely absent from information and promotional material circulated about the scheme which can be considered a full scale extension to the City of London's ring of steel. As noted in *The Observer* newspaper (9 February 2003):[37]

> Security cameras will be able to zoom in on the faces of drivers entering London's congestion charge zone as part of a sophisticated 'ring of steel' around the capital.

The Observer alleges that 'MI5, Special Branch and the Metropolitan Police began secretly developing the system in the wake of the 11 September attacks'. As such, 'the controversial charging scheme will create one of the most daunting defence systems protecting a major world city'. It is also alleged that 'the system also utilises facial recognition software which automatically identifies suspects or known criminals who enter the eight-square-mile zone', although this type of technology is unlikely to be available at present. Using facial recognition technology to 'snap' the driver rather than the number plate of a vehicle would also need necessitate different legislation under the 1998 Data Protection Act and a rigorous code of conduct set up for operators and monitors of the system.

Not surprisingly, civil libertarians feel 'misled' over this hidden use for the scheme which is promoted as an attempt to beat traffic congestion. *The Observer* article notes Gareth Crossman, policy director of Liberty who stated:

> There is an issue we are concerned about which is called 'function creep'... This is where we are told that a system is being set up and used for a certain purpose and then we find out it is being used for another totally different one. It is a dangerous precedent...We would be concerned that it would be just a 'fishing' exercise where large amounts of data are passed over to the police or the security services and they just sift through it.

Notes

1 Budd (1998) indicates that two reports in the 1990s were central in establishing the strategies of the Corporation as they attempt to keep London, and especially the City, at the forefront of the global economy. These were *The City Research Project* (1994) and *Four World Cities: a comparative Study of London, Paris, New York and Tokyo* (1996) written by Llewelyn-Davies.

2 The London Pride Prospectus (1995).

3 See for example Thornley (1998), Klienman (1999).

4 See for example *Meeting the transport needs of the City* by Glaister and Travers (1993).

5 The LBA was formed in 1983 followed by the ALA in 1986 after disagreements in the aftermath of the abolition of the GLC in 1986. These merged in 1994 to form the Association of London Governments, as a result of most boroughs becoming Labour controlled after the general election.

6 Kelly (1994a) also points out that no American bank received such a letter, perhaps related to the continuing support for the Irish republican cause in some sections of American society.

7 Cited in Buckingham (1994, p.8).

8 Corporation Police Committee minutes 24 November 1993.

9 Cited in Elliot and Mackay (1993).

10 See Jack (1993, p.6).
11 See Corporation of London 1993a and 1993b.
12 However, only 279 replies were received in relation to the first document on security initiatives and 700 for the report on the traffic management scheme (i.e. the ring of steel). This was partly put down to the consultation being undertaken in the summer holiday period (Policy and Resources meeting, 16 September 1993). This however is not a particularly valid sample given the low response rate.
13 Evaluation was also in part based on the cost of terrorism insurance, although the fact that a scheme had been established meant this became of secondary importance to businesses that needed to maintain a presence in the City.
14 The process of constructing the ring of steel was itself helped by meetings in Belfast between the RUC and the City of London Police in 1992/3 which saw the beginnings of an informal partnership approach develop between these forces. Subsequently the two forces were involved in an Exchange of Inspectors program in 1994, which allowed an exchange of ideas between the two forces to continue in relation to counter-terrorist tactics (see for example Ayres, 1994).
15 Minutes of the Corporations' Police Committee, 24 November 1993.
16 The survey was carried out by Applied Property Research who were concerned with the effect of the bombing on property prices, seen as a key measure of economic competitiveness. This report represents a telephone survey of 100 corporate occupiers in the City. The majority (79%) were related to banking (46%) and insurance (23%), with most having between 100-499 employees (43%) or 500-5000 (39%).
17 Cited in Thompson and Waters (1993)
18 Cited in *The Times*, 25 April 1994.
19 See Planning and Transportation Committee, 1 March 1994.
20 However the City of London Environmental and Amenities Trust did officially object to the scheme being made permanent on the grounds that full access had not yet been assured for cyclists. A Corporation Policy as part of the Unitary Development Plan (UDP TRANS 28) requires that pedal cyclists are given special consideration when traffic management schemes are introduced.
21 Cited in Smith (1993).
22 Cited in Smith (1993).
23 Ibid.
24 In particular Aldgate and Tower Hill gyratories have been affected, although the opening of the Limehouse link tunnel could well have affected this.
25 Minutes of the Policy and Resources Committee, 2 June 1994.
26 Corporation of London publicity leaflet, January 1997.
27 Cited in Waugh (1997).
28 See for example Planning and Transportation Committee (20 May 1997) and Policy and Resources (5 June 1997).
29 Cited in the *Daily Telegraph*, 19 September 1997.
30 See for example the *Daily Telegraph*, 12 March 1998.
31 Cited in Ford (1993).
32 This view was also prevalent in the Docklands after they introduced similar camera technology in 1996 (see Chapter 8).
33 Cited in Dillon (1996, pp.292-3).
34 Cited in Kleinman (1999, p.14).
35 London Chamber of Commerce (1995 and 1996).
36 Such a scheme is intended to reduce congestion (by 10%) which was regarded as a major negative feature for international businesses locating in central London. This was part of a wider integrated transport strategy (Greater London Assembly, 2000).

37 Townsend, M. and Harris, P. (2003) 'Security role for traffic cameras', *The Observer*, 9 February 2003.

PART III
TERRORISM, RISK AND THE FUTURE CITY

Chapter 8

Beating the Bombers: a Decade of Counter Terrorism in the City of London

Introduction

Previous chapters in this book have highlighted the impact of a number of different responses to the risk of terrorism in the City of London since the early 1990s. In so doing it has sought to address a number of key questions.

First, how was the physical form of the landscape in the City changed as a result of attempts to 'design out terrorism'? Who was responsible for such measures being implemented? How did such alterations change over time in relation to the prevailing socio-economic and political context? And, to what extent do the security changes adopted in the City relate to strategies adopted in other cities to 'design out crime'? In particular, the analogy of territoriality was adopted as a way of articulating literatures on the fortification and privatisation of urban space. This was presented in terms of how contemporary cities are enclaving and restructuring themselves in relation to capital flows, but also physical and financial risk. Furthermore, importance was placed on understanding how influential urban managers within a specific territory act as 'conditioning agents', creating an urban landscape that becomes, in the words of David Harvey (1990), 'necessarily fragmented'.

Second, this book was also concerned with the question of how the establishment of anti-terrorism security measures in the City was related to a distinctive set of local governance arrangements operating within, and in relation to, the Square Mile. This analysed the extent to which the 'institutional thickness' of the City helped the Corporation of London to establish the ring of steel. This aspect of the enquiry also sought to explain how these institutional networks were themselves modified as a result of the continuing terrorist risk. In addition, the relationships between the City and central government were investigated in relation to how the terrorist threat, and the subsequent construction of the ring of steel, affected the promotion of London on the global stage.

Third, this book has been concerned with the impact of terrorist risk in the City in a financial sense, and has investigated how insurers attempted to redistribute the financial risk of terrorism away from the Square Mile. It has also sought to illustrate the regulatory role played by the insurance industry in shaping the landscape of the City through their policies of encouraging the retrofitting of security measures and the development of enhanced risk management procedures. Finally, this book has sought to relate the experiences of the City of London anti-terrorism measures to strategies that other cities have adopted in their fight against terrorism in the post 9/11 era.

This chapter is concerned with linking the City's response to terrorism with wider conceptual ideas within the social sciences. First, it will highlight the dangers of attempting to replicate the strategies that have developed in the City in other locations arguing that what developed to counter the terrorist threat in the Square Mile is geographically and historically specific. This will be exemplified by cases from the London Docklands and central Manchester. The second part of the chapter will highlight attempts to design out terrorism, whist the third section will focus upon the importance of a strong institutional structure in enabling the City to construct anti-terrorist defences relatively unhindered. The final part of the chapter will draw on the analogy of territoriality to highlight the impact of risk and risk aversion in constructing a urban landscape of 'wild' and 'tame' zones.

Challenging the 'Normative Drift'

Within the social sciences the events that occurred on 9/11 have led to many discussions about the appropriateness of current interpretative techniques especially in relation to the form and functioning of urban areas. For example Kellner (2002) argued that the shock caused by 9/11 suggested that normative social theory or models can no longer explain the full complexities of the situation and that new models are required to account for the complex and abstract notions that we are now faced with. In particular he argued that we 'need to show how events like September 11 produce novel historical configurations, while articulating changes and continuities in the present situation' (p.149).

Kellner's assertion links too much recent urban research where there has been a tendency to assume that the increasingly prevalent trends of urban fortification and the development of new forms of entrepreneurial urban governance can provide a model for the design and functioning of the future city. It is often argued that such a framework will result in a safer, increasingly egalitarian, and economically prosperous society. However, such accounts have a tendency to generalise their findings proclaiming to represent the future of urbanism without accounting adequately for local circumstances of place. This represents part of a much wider 'normative drift' within the social sciences, where new models and conceptual approaches, to represent new realities, are commonly sought given a new set of contemporary social, economic and political conditions which cannot be explained within traditional explanatory frameworks.

Such normative accounts have tended to dominate many recent studies in urban theory. These in particular have related to the differences, and shift between: modernism and postmodernism; managerialism and entrepreneurialism; government and governance; industrial society and risk society; and Fordism and post-Fordism. This has meant that research that adequately assesses the complex realities of cities is rare, with much attention being paid to the 'big picture' of urban restructuring due to the large-scale social, economic, political and technological transformations that are being portrayed within the social sciences. In urban studies, many of these broad conceptions were based upon the development of LA as the archetypal postmodern city. More recently, in the aftermath of 9/11 some commentators have argued that it 'all came together in New York' (Catterall, 2002) and that the measures taken in response will have a lasting impact upon global urbanism.

The findings presented in this book seek to challenge this assumption and have argued that the events that occurred as a result of the risk of terrorism in the City of London were a result of a unique set of historical, geographical and institutional factors which should not be mapped uncritically to other areas. Indeed, as shown below attempts to set up City of London style 'rings of steel' in other parts of the UK have been played out in very different ways.

The Mini Ring of Steel

Another site in London that was the focus for counter-terrorist planning through the 1990s was the London Docklands and in particular the Canary Wharf complex (see Figure 8.1). This area was subject to a failed terrorist bombing in 1992 when the vehicle containing the bomb was spotted before it exploded under the main Canary Wharf Tower, as well as a devastating explosion in the southern part of the Docklands in February 1996. After the 1996 bomb the powerful business community successfully lobbied the Metropolitan Police to set up their own anti-terrorist security cordon. However, the beginnings of a coordinated security response to terrorism began in the wake of the bomb scare in 1992.

Following the 1992 Canary Wharf attack an area-based security liaison group was set up which incorporated representatives from Canary Wharf, London Docklands Development Corporation (LDDC), Tower Hamlets Borough Council, the Metropolitan Police, and the Docklands Light Railway. Developing a security strategy for this area was not an easy task given the unique nature of the Docklands area which, unlike the City of London encompasses a range of functional uses – public housing estates, private residential developments and leisure facilities as well as private commerce.

Whereas the entire London Docklands was considered at risk from terrorism, particular concern was expressed about the susceptibility of Canary Wharf to terrorism attack given its iconic status. As such managers at Canary Wharf initiated there own mini-*ring of steel* essentially shutting down access to 'their' estate (Coaffee, 1996a, 1997, 2001; Graham and Marvin, 2001). Such an approach combined attempts to design out terrorism with changing approaches adopted by the police and private security industry. Security barriers were thrown across the road into and out of the complex, no-parking zones were implemented, a plethora of CCTV cameras were installed and identity card schemes were initiated (Coaffee, 1996b, 1996c). Such images are shown in Figure 8.2.

Figure 8.1 The Canary Wharf Tower in the London Docklands

Figure 8.2 Security features at the entry to the Canary Wharf complex

Comparisons were made at this time between the City of London and Canary Wharf in terms of security procedures. As a Business Director in Docklands commented 'nothing goes into and out of Canary Wharf or the other adjacent quays… without the scrutiny of a vigilant security operation' (Peraticos, 1993, p.17). He continues:

> Security is an intrinsic part of the design of Docklands. Close-circuit video cameras network the area, wide open spaces expose and preclude suspicious activity...Irrefutable proof that the system works was offered when a van packed with explosives outside Canary Wharf after midnight was spotted immediately, thereby avoiding loss of life and disruption. There are many secure buildings in the City. However the lesson must be learnt that this is not enough. The whole area must be secure – Docklands is such a place.

Outside of the Canary Wharf security scheme, the LDDC, through liaison with their private security advisors also deployed a number of foot patrols in the area as well as at other vulnerable target locations such as the London City Airport and the Royal Docks. These patrols subsequently reduced or intensified in relation to the perception of the terrorist threat. At this time there were also calls for the LDDC to set up a centralised CCTV scheme. However institutional problems meant this was never feasible. These were mainly linked to the proposed cost and the short life span of the LDDC which came to an end on in March 1998.

An Iron Collar for the Docklands?

At 7:01pm on 9 February 1996 a large vehicle-borne explosive device left by the Provisional IRA exploded near South Quay railway station in the London Docklands.

The bomb immediately began a process of security reassessment with proposals for radical security enhancement, modelled on the City of London, developed to form an 'Iron Collar' for the Docklands.[1] Similar to previous bombings that had occurred in the City, there were fears that high-profile businesses might be tempted to relocate away from the Docklands. As the LDDC chief executive noted in the *Daily Telegraph* 'the scheme is meant to act as a deterrent and make business and residents feel that measures are in place to prevent a repeat of last month's bomb'.[2]

Subsequently a security cordon was initiated for the Isle of Dogs (see Figure 8.3). This comprised four entry points which at times of high-risk assessment would have armed guards. High resolution CCTV cameras were also installed. The system was devised so that control of the access points could be manipulated so that Police could undertake physical searches of vehicles entering the cordon rather than relying solely on the electronic recording of the driver and vehicle. The scheme was much easier to devise than the ring of steel in the City given the small number of entry and exit roads onto the Isle of Dogs.

The scheme was a jointly funded initiative between the London borough of Tower Hamlets, Canary Wharf Management, LDDC, and Docklands Light Railway. The Metropolitan Police agreed to manage the system when it was installed.

In November 1996 a three week trial period for the scheme passed without any major problems. Like the City of London's ring of steel there were some objections with regard to traffic problems and civil liberties and it was argued by local Borough Council officials 'that extensive consultation would be needed before measures were finally approved – if you are looking at cordoning off the Isle of Dogs you are looking at 11,000 families'.[3] Furthermore, this was compared to the City of London:

> There is a civil liberties aspect to it. Very few people live in the City, so their ring of steel is a completely different ball game from having your picture taken by security cameras day in day out as you go to and from your home.[4]

However, in general there was unanimous support for such a scheme. This opinion was expressed at a series of public meetings where plans for what the LDDC called the *ring of confidence* were suggested (Coaffee 2000b). For example the Chief operating officer at Canary Wharf, Gerald Rothman indicated at this time that the South Quay bomb had only minor impact on the success of the overall Docklands development. He pointed to the fact that London as a whole is a potential target when indicating that the Isle of Dogs as a whole and specifically Canary Wharf now has highly advanced security systems, noting that 'tenants appreciate the level of security they enjoy'.[5]

The most noticeable difference between the scheme initiated in the Docklands and that in the City was the overt advertising of the Docklands security cordon. Figure 8.3 shows one of the large signs at entry points into the cordon. As previously noted in the City of London the ring of steel was not advertised as a security cordon, rather it was promoted as a traffic and environment scheme.

Whereas the City of London and the London Docklands were able to initiate a counter-terrorist scheme relatively easily given the perceived risk of attack and its powerful business community, other urban areas in the UK did not find it so easy. On June 15 1996 when the Provisional IRA exploded a 3,300lb bomb outside the Arndale shopping centre in Manchester. This led to the Manchester business community calling for enhanced anti-terrorist measures to secure the city's commercial heart against further bomb attacks. As such the some members of the Manchester business community called for the City of London-style security cordon to be installed. However upon closer inspection it was not considered viable to implement such a cordon in Manchester:

> Calls for a ring of steel were an initial reaction and given the compact nature of the city centre it was considered that it was not the most appropriate measure. A City of London system with entry points is unlikely due to cost. The City of London has a very powerful business community, which helps. That is not saying Manchester doesn't, but not to the same extent.
>
> (Senior Greater Manchester Police Officer)

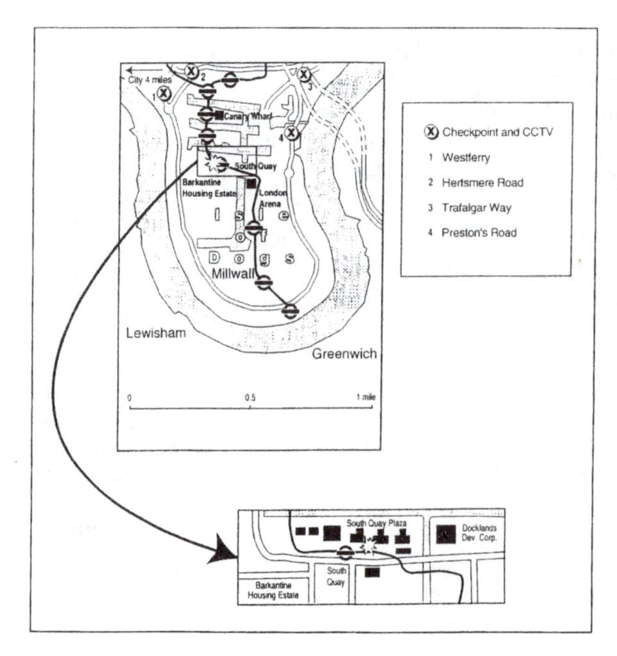

Figure 8.3 The Docklands Iron Collar

Designing Out Terrorism in the City

The obsession with urban security has in recent years led architects and planners to design buildings and spaces that are infused with notions of defence and security in residential, retail, leisure and commercial areas according to the preferences of wealthier residents, consumers or businesses. Examples, especially from the United States, but now increasingly from other countries, indicate that 'fortress urbanism' with its walls, security guards, pedestrian partitions, traffic barricades, gates, cameras and other physical measures are now *de rigueur* for creating well defined pockets of urban civility, community and mutual support within an urban landscape that is often perceived as dangerous or unsafe. Indeed, the 'fortress city' approach can be seen as symptomatic of the so-called postmodern city, or 'New Urbanism' which has a habit of 'privileging spatial form over social processes' (Harvey 1997, p.2).[6] Such design-led approaches are seen by many to contribute and actively reinforce the segregation and fragmentation of the urban landscape. As Lehrer and Milgrom (1996, p.15) noted, 'the reification of physical models is used by the architects of New Urbanism as a strategy to create local community, by reproducing a physical environment that fosters greater causal social contact within the neighbourhood. However, these architects fail to sufficiently consider segregation within the greater urban area…and may, in fact, perpetuate it'.[7]

This is a similar argument to that used against the ideas of defensible space developed by Oscar Newman in the early 1970s, in which he argued that safer neighbourhoods could be achieved through re-design. Importantly in the post 9/11 context, Newman's ideas are now seen as a cornerstone of new urban policies in the USA, and increasingly in the UK and other Western countries to reconfigure existing neighbourhoods, design new urban spaces and residences and to reduce the risk of terrorism (Vidler, 2001a, 2001b).

Attempts to 'design out terrorism' in the City of London since the early 1990s exhibit similarities with, as well as differences to, the fortified and fragmented model of future urbanism that many recent accounts of cities portray. As a result of the terrorist threat, the landscape of the Square Mile was significantly altered as attempts were made to reduce the City's vulnerability to further terrorist strikes. As a result, previous methods used to 'design out crime' were adapted and modified in relation to the threat of terrorism. These changes aimed to provide a secure and attractive place in which the City could grow and compete globally. In particular, these defensive strategies were all based, to varying degrees, on access restriction, enhanced electronic surveillance and the risk management techniques adopted by the agencies of security.

Within the City, the spatial emphasis of such security approaches served to reinforce the control of space in specific ways. As such, there was a gradual enhancement of security, both in terms of the number and concentration of strategies utilised, as well as in the ever-increasing spatial area where such strategies were focused. Alongside the ring of steel, territorial control was also reinforced at the level of the individual building (or in some cases groups of properties). This occurred through the addition of security measures in order to 'harden' the landscape, to reduce the damage that might be caused by further bombing. This was facilitated by insurance premium discounts for the retrofitting of such security measures.

Subsequently, as a result of enhanced security strategies the City was perceived by many to be a fortress, a *cordon sanitaire*, intended to produce a safe working environment. As one commentator put it just after the ring of steel was set up:

> Motorists attempting to cross the City are getting used to closed access routes and diversions...and few would be surprised if they were stopped at a road block by a policeman toting a sub-machine gun: welcome to Fortress London, 1993.[8]

However this view is not an accurate reflection of the situation that developed and is infused with the rhetoric emanating from American cities where the territorial control of space is seen first and foremost in terms of the division and segregation of the urban landscape in to mutually hostile units. The 'fortress London' idea was also a reflection of the parallels that were frequently drawn between the City's attempts to deter terrorism and those employed in 1970s Belfast where a security cordon was set up with high steel gates encircling the central city area.

The City's ring of steel in this sense was far more *symbolic* as it did not rely on overt fortressing approaches. Consequently, the ring of steel represented a highly visible deterrent, allowing an apparently safe and secure place in which to conduct business to emerge without high levels of fortified security having to become part of the physical landscape. This approach exemplifies a less well-documented notion of territoriality namely in terms of a 'landscape of symbols' where territorial boundaries are reinforced through the manipulation of space and landscape artefacts within an area. In the City's situation this involved measures such as such as road closures, police checkpoints, pedestrianisation and tactically placed CCTV. In short, altering the spatial configuration of the built environment developed the impression of a highly defended landscape employed by, or on behalf of, powerful groups in an effort to create territoriality.

This enquiry has also illuminated another important aspect of the fortress urbanism approach, namely the assumption that increased physical security gives increased feelings of safety. In recent years a number of pieces of urban research have increasingly suggested that fortified landscapes, which are designed to reduce fear, can in fact exacerbate the fear of living, or working, in a particular part of a city (see for example Ellin, 1997). In the context of this book, this relationship between fear and form was of importance especially after the 1993 bomb where there were calls for Belfast-style security to be adopted. Despite the development of a powerful pro-security discourse, the need for proactive security enhancement was tempered by the realisation that radical security measures could be detrimental to the image of the area. As a result, the decision was taken not to construct a militarised landscape especially given the efforts of both the City and Central Government at this time to promote London internationally. Despite this decision there were some complaints, especially during the ceasefire period, that the ring of steel was still a powerful symbol of the risk of terrorism and that it should be completely removed. However, this view soon changed after the Docklands bomb when most businesses in the City were very much in favour of the ring of steel staying and being extended, coupled with further defensive adaptations to the landscape to deter the terrorist threat.

Importantly though, despite the pro-security agenda pushing forward the construction and subsequent expansion of the ring of steel, the Corporation, for both legal reasons, and in an attempt to remove anti-terrorist references from the City's landscape, continually attempted to justify the ring of steel in relation to unintended improvements in traffic flow, reductions in atmospheric pollutants and falling levels of crime. As previously noted, the ring of steel was initially referred to as the 'experimental traffic scheme' and prominence was given to the opportunities the terrorist threat had inadvertently presented to the City in terms of enhancing the environmental quality of the street scene. For example, the Corporation noted in November 1993 in relation to the development of the ring of steel that 'great care will need to be taken with the works that are to be carried out where they impact on conservation areas and listed buildings and their settings, to ensure that the townscape in general, and these aspects in particular, are enhanced'.[9] Later, after the Docklands bomb, the ring of steel was officially referred to as the 'traffic and environment zone' in further recognition of the reductions in traffic flow and pollution that occurred, and which, it was proposed, could be extended to cover more of the Square Mile.

Strengthening the City's Position through New Institutionalism

As with the 'fortress city' and New Urbanism approaches, normative assumptions are also widespread in the field of urban governance. As such, a number of commentators have drawn attention to the dangers of uncritically using terms like 'institutional thickness' and 'institutional capacity' assuming that economic competitiveness is necessarily dependent upon successful partnerships and, that these relationships are politically neutral. Whilst inter-agency co-operation is often an important facet of developing a 'competitive edge', institutional relations also reflect broader urban power relations. In particular, it has been argued that within the development and deployment of urban policy agendas certain elite-driven 'voices' are more dominant than others. This represents a dangerous precedent for urban affairs as stakeholders increasingly attempt to hold down global processes at a local level. Often this results in the search for consensus blinding urban policy makers to alternative ways of developing cities, which are advocated by coalitions of different and less powerful stakeholders. This book has exemplified a number of such aspects of contemporary governance agendas: the merging of local and global contexts of action; strategic collaboration between key stakeholders to achieve mutual benefit; and the marginalisation of less powerful 'voices' by dominant discourses.

It has been argued that the City's anti-terrorism effort was a response not only to the localised risk of terrorism against the Square Mile but also a reaction to London-wide agendas which were being advanced by Central Government. In the early 1990s the increasing business involvement in the place promotion of London, and in particular the place of the City within this overall schema, meant that that the Corporation of London, often through the work of Michael Cassidy the chairman of its Policy and Resources Committee, began to get involved with London-wide initiatives and partnerships for the first time in recent decades. The City began to establish new institutional relationships with the neighbouring boroughs, as well as Central Government, as the importance of promoting London as a world city began to permeate all aspects of London governance. Thus when the ring of steel was suggested, it was promoted not just as increasing security for the City but as beneficial to the reputation of London as a whole in the wake of terrorist attacks.

Given the broader geographic context of economic global terrorism in the early-mid 1990s, London was not seen as being at greater risk than its economic rivals. Indeed, the ring of steel became a very visible symbol that in the City, the agencies of security and the political authorities were actively responding to the threat. The perception the City tried to convey at this time was that they were better prepared than their economic competitors to both deter bombings or to cope with the aftermath of a successful terrorist strike.

The anti-terrorism measures put in place, far from decreasing the City's attractiveness as a financial centre as the Provisional IRA had hoped, in fact meant that the City could maintain and enhance its global economic standing. Indeed, it was not long after the ring of steel was constructed that the landscape changes that occurred in the City were shown as improving the quality of life in place promotional material for London. Given concerns in many of the world's larger cities that terrorism could adversely affect their attractiveness to inward investment, the London Chamber of Commerce's 1995 and 1996 *Invest in London* publications both highlighted the ring of steel as being a key contributor to the enhanced feeling of safety in London, thus leading to an increase in 'business confidence'. As a spokesman for the London Chamber of Commerce further stated 'we feel safe inside the cordon and want to feel part of its inner sanctum'.[10] However, consequently, some viewed the ring of steel as purely a public relations exercise that aimed to increase the attractiveness of the City and limit the potential exodus of international firms who were fearful of further terrorist attacks. This trend continues today with high levels of anti-terrorist security displayed by the City being highlighted in London-wide business investment brochures.

Whilst the construction of the ring of steel in the 1990s undoubtedly allowed London to continue its 'internationalisation',[11] the ever more cosmopolitan business community also made the Square Mile an increasingly attractive terrorist target, not just for the Provisional IRA, but for other terrorist groups. This illustrated how political conflicts world-wide were capable of being played out in the City due to its array of foreign firms and the ability of the mass media to publicise such attacks. For example, the two main bombs in the City in 1992 and 1993 were not just targeted at symbols of the British establishment such as the Bank of England, but at foreign investors, especially the Japanese. The tactics of the Provisional IRA at this time were directly related to the underlying political climate, and aimed to cause relocation of these businesses away from areas of perceived danger. They aimed to undermine confidence in the Square Mile's reputation as a safe international business centre, in particular by threatening its position within European finance. In short, the economy dictated the target, whilst the political situation determined the timing of attack. The construction of the ring of steel was the response.

Strategic Collaboration and the Ring of Steel as a Negotiated Statement

During the late 1980s and early 1990s the City's business community, and in particular the Corporation of London, began to take a lead role within the new governance structures of London. This involved increasingly entrepreneurial approaches to obtaining inward investment as well as the development of institutional links both within the City, and between the Corporation, the London boroughs and, central government.

The mobilisation of anti-terrorist strategies in the City therefore illuminated the increasing significance of institutional context within local decision-making. In particular, it showed that a shared discourse and strategic collaboration amongst institutions and organisations is often central to urban policy and planning decisions. With regard to the setting up of the ring of steel, liaison between a number of stakeholders occurred – the police and private security personnel, planners, designers, politicians, business interests and environmentalists – about to how to best design a landscape that would deter terrorism, whilst maintaining and enhancing the attractiveness of the City as a place of business. The development, and maintenance, of a strong local 'institutional thickness', co-ordinated through the Corporation of London, was essential in creating the 'capacity' to develop the ring of steel quickly, and with minimal interference from outside influences.

Localisation in this context was seen as an institutional strategy which attempted, through the shared frames of reference of the key urban stakeholders, to maintain the City's global economic position.[12] This capacity building potential was based on a number of principles. First, the unanimous financial and political backing of the Corporation of London, second co-ordination between the police, businesses and the Corporation, and thirdly, 'political will' as encompassed in the dynamism and vision of certain key urban managers.

Of particular note is the contribution of Michael Cassidy, the chairman of the Corporation's Policy and Resources committee, during the mid-late 1990.[13] Cassidy became the mouth-piece for the City with regard to anti-terrorism security, but his contribution should be seen within the wider context of his role in London-wide agendas. As Hebbert (1998, p.121) noted, Cassidy 'became the first City councilman within historical memory to acquire a London-wide political reputation. Without his statesmanship the Corporation might not have shaken off its introversion, and London's tissue of collaboration and partnership would have looked much weaker'. In relation to anti-terrorism, Cassidy was the chief co-ordinator of the pressure put on the government to get involved with terrorism insurance and organised the City's response to the second bombing in 1993, liaising with business communities and 'brokering deals' with neighbouring borough councils and central government to allow the ring of steel to develop.[14]

However, the changes in the physical form of the landscape which occurred as a result of the threat of terrorism, and the implications for the economic competitiveness of the City, were not just down to the Corporation of London. This research has shown that a number of divergent coping strategies were adopted by different institutional groupings. This occurred within an overall context of strategic collaboration in support of security enhancement.

The subsequent form and function of the ring of steel can therefore be seen as a negotiation between a number of powerful interests. This evokes Pahl's (1970) urban managerialist notion that the built environment evolves as a result of conflicts between key urban managers with different degrees of power to shape socio-spatial outcomes. In particular, the different strategies of the Corporation of London (localisation for globalisation), the City Police (the territorial control of space) and the insurance industry (risk spreading) led to the Square Mile being viewed in different ways – as a 'reputational area', a terrorist target to be defended, and a risk hotspot that requires 'cooling' through risk and insurance distribution. The resulting security measures were in large part a result of the negotiation between these strategies. In addition, considerations of heritage, cost, legality, and, public opinion also served to affect how the ring of steel subsequently developed.

Negotiation was also required to deal with the associated geographical effects related to the displacement of the risk of terrorism and enhanced traffic flows to areas around the edges of the ring of steel. Within the City, complaints from prominent businesses led to reassurances from the City Police that additional policing would take place in the so-called 'collar zone'.[15] When similar complaints were made from the neighbouring boroughs, or from civil liberty groups, little change or accommodation was forthcoming from the Corporation of London.

Outside discourses that emerged to challenge the powerful City position through the statutory consultation process, for a while at least, had the potential to undermine the wishes of the Corporation of London to construct a permanent ring of steel. Ultimately, however, negotiation between these interest groups, the Corporation and Central Government, meant that opposing voices had little impact on the overall construction and extension of the ring of steel and the associated security initiatives employed. In particular, the City was accused of acting paternalistically and of only paying 'lip-service' to consultation about ideas that it had effectively already decided to implement.

This, like much research before it, exemplified the increasingly apparent trend in Western cities for the agendas of local governance to be controlled by dominant elite of businesses and political interests, which recognises, but then marginalises, other contesting discourses. In the City this conflict of interest was articulated through the distinction between the 'powerful' inside discourse and the 'powerless' outside discourse. Consequently, this highlighted how ultimately, diversity of meaning was minimised in favour of dominant and powerful representations about the actions and outcomes required.

Risk, Security and the Protectionist Reflex

Findings from this research exemplify how the concept of risk has the power to shape the form and meaning of landscapes. In particular, risk theory has also shown how insurance mechanisms increasingly attempt to distribute risk and contribute to the prevalence of fortified risk-adverse landscapes, often as a result of 'insurance redlining' in what Ulrich Beck referred to as 'loser regions'. Indeed, Beck in a recent book *World Risk Society* (1999, p.153), argued that in such vulnerable areas a 'protectionist reflex' is evoked where a 'withdrawal into the safe haven of territoriality becomes an intense temptation'.

It has been shown that whilst the ring of steel was seen as a 'ring of confidence' by many businesses, financial security in the form of terrorism insurance was also vital in maintaining the profile of the City in the global market place. This was conceptualised in six distinct stages in Chapter 6: emergent risk; reflecting on local terrorist risk; transferring the risk nationally; concentrating the risk in the City; fluctuating risk and alternative risk sharing mechanisms; and the globalisation of risk. These stages showed the different ways in which the UK insurance industry attempted to distribute the financial risk of further bombings away from the City – first, through the threat of withdrawing from the market; second, through the pressure wave insurers helped develop, in collaboration with the Corporation of London, to get the Government to underwrite terrorism insurance. This occurred amidst predictions that the insurability of the City would further jeopardise the recession-hit economy. It was feared this would make the City an unattractive area in terms of occupancy and investment. Subsequently, the zoning policies the insurance industry helped establish, attempted to distribute financial risk away from the City whilst in reality creating a risk hot-spot centred on the Square Mile. The dangers of economic exposure therefore became one of the catalysts for the ring of steel, which was seen as a vital risk management measure, allowing City businesses to remain insurable.

As a result of terrorist risk the insurance industry began to have an increasing influence as a regulator of the urban landscape of the City. They did this by offering significant premium reductions on terrorism policies for the retrofitting of security devices and the development of advanced crisis recovery and security planning. The measures encouraged by the insurers were important in reinforcing the centrally organised security strategies of the police and Corporation of London, as well as contributing to the overall culture of security within the Square Mile.

This enquiry has also illuminated a number of facets related to the changing way in which risk is managed within contemporary society, and in particular, how risk management is now embedded into social and institutional structures. Whereas in previous eras the police have been seen as the sole agent of social control, in recent years increased institutional demands for risk-related knowledge have occurred, most notably from insurers, who have been under increased economic pressure to minimise liability. Subsequently, judging risk is now undertaken by a plethora of institutions, all acting in their own interests and all focusing on the fear of 'bad' risks. In this sense Beck (1999, p.16) argued that 'risk-sharing' can increasingly become a 'powerful basis of community', which can have 'territorial aspects' – 'risk-sharing further involves the taking of responsibility, which again implies conventions and boundaries around a risk community that shares the burden'. In particular, it has been highlighted in previous chapters how different agencies had different 'ways of seeing' terrorist risk and employed different strategies to reduce the threat whilst still being part of this risk community in the City.

Given the City of London's financial position within global markets, inevitable questions have been asked after 9/11 regarding how the Square Mile would cope with a similar attack. Views appear to be varied as to the extent to which financial institutions are developing adequate disaster recovery or contingency plans. Some claim that it would be 'business as usual almost immediately. As a London Stock Exchange (LSE) spokesperson noted on BBC news a year after 9/11:

> Put simply, if disaster were to strike, City trading should continue without interruption.

LSE however have, and have had for a number of years, a 'hotsite' – a fully sourced back up office from which all electronic trading could continue if there main headquarters in the centre of the City was damaged in a terror attack. Compare this with the situation in New York where no such back up was available to financial corporations pre 9/11. The UK's Financial Services Authority have also highlighted that all of the UK's 35 biggest financial institutions have systems which will allow them to trade if they are forced to relocate away from their main centre of operations. However, others are more sceptical. Despite the Corporation's Security and Contingency Planning Group assisting businesses in the City with the development and exercising of their business continuity plans and appropriate disaster response preparation, it was estimated that between 30-40% of financial institutions have no such plans in place. Furthermore, at an area level the Corporation of London is involved with running disaster response scenarios both within the Square Mile and in coordination with London as a whole. These scenarios are rapidly different given the changing nature of the terrorist threat:

> Every contingency plan has had to be measured against a new higher level of disaster. The worse case scenario is no longer a two thousand pound bomb going off.
>
> (Corporation of London spokesman)[16]

He however continues by noting that:

> Although those charged with disaster coordination are making confident noises that preparedness is a lot higher than in New York on September 11, the City still has a long way to go... We are in a different world – planning has to take account of truly massive events – so far so good but more still needs to be done.

Dislocating the City from London

The overall spatial imprinting of risk, defence and new governance agendas in the City of London led to a juxtaposition of landscapes of power with defence. As Urry (2002, p.64) highlighted the ring of steel provides a 'physical and symbolic separation' between 'wild' and 'tame' zones, developing a powerful 'city of control'. As such, the creation of the City's anti-terrorism defences should not be viewed purely in terms of security but contextualised within the more extensive transformations that occurred within the City's urban landscape at this time, related to its global positioning.

The City is an area that displayed its power globally through its built form, as well as locally, through an asymmetry of power with the neighbouring boroughs, creating what Sharon Zukin termed a 'landscape of power'. This enquiry has shown that in the City, notions of power and defence formed an uneasy, yet essential, relationship, as displays of economic power increasingly incorporated defensive notions of surveillance and social control. The ring of steel symbolically and functionally served to institutionalise further the separation of the City from its neighbours, through the physical imposition of advanced security measures and access restrictions. This formed what was previously referred to in as a 'landscape of defence', which provides a reliable background to everyday life through formal and informal defensive strategies enacted for a variety of social, economic and political reasons. However, as has also been noted, the City historically has always displayed such notions of power and defence, and in this context, the response to the terrorist threat should be seen as the latest attempt by the City to defend itself physically, as well as economically. In particular, the encircling Roman defensives and the mediaeval system of walls and gates, which were in place until the eighteenth century, served as historical antecedents of the ring of steel.[17]

The impact of landscapes of power and defence can be graphically illustrated by the contrasts between the built form and economic power of the City, and its surrounding areas, which are often characterised by low-income housing. In the mid-1990s, the City, surrounded by the ring of steel, was situated cheek by jowl with some of the poorest and marginalised communities in Britain, such as Southwark and Tower Hamlets.[18] This illustrated the paradoxical condition of increased global connectivity alongside increased local disconnection – a condition that was reinforced by the ring of steel, which exacerbated this boundedness between the City and its surrounds. In short, the ring of steel can be seen as an attempt to both physically and symbolically keep the City in the City.

Interestingly, the development of the ring of steel and its impact in fragmenting the centre of London occurred at the same time as the institutional relationships between the City and the boroughs had never been better with new institutional networks and collaborative efforts being developed between the local authorities in central London. On the one hand therefore, the City was engaging with the London boroughs for the mutual benefits it could bring to London, whilst on the other hand it was disengaging and protecting itself without consideration of the knock-on effect on other areas.

The next chapter will highlight the response of the City Police and the Corporation of London to the events of 9/11 as well as critically analysing many of the discourses surrounding the future of cities *per se* that this unprecedented event has generated.

Notes

1 Indeed the same firm of engineers were employed as in the City were approached to set up the scheme.

2 12 March.

3 *Electronic Telegraph*, 12 March 1996.

4 Cited in *The Times*, 12 March 1996.

5 Cited in *Investors Chronicle*, 5 July 1996.
6 Cited in Fainstein (2000).
7 Ibid.
8 See Pratt (1993, p.20).
9 See Minutes of the Policy and Resources committee 11 November 1993.
10 Cited in Gusmaroli (1993, p.1).
11 See Jacobs (1993).
12 It contained more than 561 foreign banks, 170 foreign security houses and 185 corporate headquarters. It also contains 37% of the largest corporate headquarters in Europe (Corporation of London, Economic Development Unit, 1997).
13 Prior to this he had been the chair of the Corporation's planning committee and was instrumental in giving permission for the building of much of the new office stock (including Broadgate) in the late 1980s and early 1990s.
14 See also Ashworth (1996).
15 Furthermore, in 1996 the Corporation, given the reorganisation of local authority boundaries, agreed to extend the ring of steel to cover much more of the City. The changes to the structure and policing of the ring of steel boundary were, however, internal adaptations within the confines of the Square Mile.
16 Cited by *BBC News Online*, 10 September 2002
17 The Roman fortification were established in around 200AD and stayed intact until about 400 AD, whilst the medieval defences were established around 1200 and remained for over 500 years.
18 This of course is not a new feature of cities with similar analogies, relating to the tendency of the built environment to fragment into pockets of rich and poor, being commonly made about Western cities.

Chapter 9

Terrorism and Future Urbanism

Introduction

In the post 9/11 era it is generally accepted that the nature of potential terrorist threats has changed, requiring alteration in counter responses from Governments and agencies of security at all spatial scales from defence of the local area to the developments of global coalitions to fight the 'war of terrorism'. Not only were the attacks against New York and Washington in September 2001 unprecedented in terms of style of attack, damage caused and insurance claims, they also brought to the fore wider concerns about different types of 'postmodern terrorism' (Laqueur, 1996) highlighting the links between new forms of urbanism, strategically targeted terrorism (especially those using chemical, biological or nuclear products) and military threat response technology.

The last chapter noted the traditional reaction to terrorism has commonly been to evoke territorial metaphors of security and safety in order to separate 'wild' and 'safe' zones. This chapter will broaden this argument and highlighted the importance of 9/11 for our understanding of the enormity of the risks that global society faces and how such mega-terrorist risks and responses have become essentially 'deterritorialised' with security implications which spread far beyond the territorial scale. Whereas previously the tendency amongst those who were, or feared they were going to be attacked, was to evoke protectionist reflexes based on securing a particular area, today such securitisation has spread to incorporate international co-operation and attempts at generating global consensus for tackling the global terrorist threat..

This chapter will also draw together accounts from a variety of academic traditions to critically highlight the rethinking of urbanism in the post 9/11 era. This will set up a threat-response continuum which at one end has 'soft' approaches relying on symbolic gestures, a culture of vigilance and an enhanced sense of community, whilst at the other end the possibility of *urbicide* – essentially the pre-emptive and the deliberate destroying of areas of cities to aid counter terrorist strategies. The chapter will conclude by summarising the key impacts of 9/11 on the City of London arguing that over the last decade this area, aided by specific morphological, institutional and technological advantages has successfully balanced the needs of security with the orderly flow of commerce.

Deterritorialising Risk Since 9/11

As noted in the previous section, pre-9/11 terrorist incidents regularly provoked a territorial security response leading to a separation between supposedly 'safe and 'unsafe' areas. Whereas the events of 9/11 provoked similar responses from agencies of security and urban managers, there was also a stark realisation that traditional counter-terrorism mechanisms are insufficient to deal with the new terrorist threats. Zygmunt Bauman (2002) for example highlighted the impact of 9/11 as the 'symbolic end to an era of space' (p.81). He argued that until recently security had also been inherently linked to territorial boundedness and the power to influence access into a particular space or area. Despite the realisation that territorial borders, be they to spaces or nation states, have been considered less important for a long time, he argued it was only with 9/11 that this became crystal clear to everyone:

> This is all over now and has been for some considerable time – but that it is indeed over has become dazzlingly evident only since 11 September. The events of 11 September made it obvious that no one can any longer cut themselves off from the rest of the world (p.82).

He continued:

> Annihilation of the protective capacity of space is a double-edged sword: no one can hide from the blows, and blows can be plotted from however enormous a distance. Places no longer protect, however strongly they are armed or fortified, nor do they give foolproof advantage to their occupiers. Strength and weakness, threat and security have become now, essentially extraterritorial issues that evade territorial solutions (p.82).

In a similar vein John Urry (2002) has also argued that 9/11 showed that 'globalization is never complete' and that 'it is disordered, full of paradox and the unexpected and of irreversible and juxtaposed complexity' (p.58). Indeed he argued that the linear conception of scale from local to global should be replaced by a concept of 'complex mobile connection', where 'wild and safe zones' have become 'highly proximate through the curvatures in space-time'.

The concepts articulated by Bauman and Urry are further enriched by Ulrich Beck who has articulated the 9/11 events in terms of newly emerging aspects of his world risk society thesis – 'It is not a matter of the increase, but rather of the de-bounding of uncontrollable risk' (Beck, 2002, p.41). Indeed he highlights this in terms of a universalising of fear of terrorist attack:

> Terrorism operating on a global scale has opened a new chapter in world risk society. A clear distinction must be made between the attack itself and the terrorist threat which becomes universal as a result of it. What is politically crucial is ultimately not the risk itself but the perception of the risk. What men fear to be real is real in its consequences – fear creates its own reality.
>
> (Beck, 2001)

Beck (2002, p.46) further argues, like Bauman and Urry, that as such 'national security is no longer national security as borders...have been overthrown'. National security is therefore seen in terms of transnational co-operation.

> Uncontrollable risks must be understood as not being linked to place, that is they are difficult to impute to a particular agent and can hardly be controlled on the level of the nation state.
>
> (ibid. p.50)

In this sense he argues that global terrorism has actually accelerated globalisation through the globalisation of politics, and that the central tenant of neo-liberal economics – that the role of the nation state will reduce and that economics will become more influential than politics – is 'exposed' by 9/11:

> The terrorist attacks on America were the Chernobyl of globalization. Just as the Russian disaster undermined our faith in nuclear energy, so September 11th exposes the false promise of neo-liberalism (p.47).

He further adds that 'suddenly the necessity of statehood, the counter principle of neoliberalism is omnipresent' (p.48). Beck argues that in order for states to defend themselves from global terrorism then they must 'denationalize and transnationalize' themselves' (p.48). The inherent danger Beck points out with this is that 'surveillance states' will be generated which:

> Threaten to use the new power of cooperation to build themselves into fortress states, in which security and military concerns will loom large and freedom and democracy will shrink' and in particular 'attempt to construct a western citadel against the numinous Other (p.49).

As such he calls for the antithesis of such surveillance states – the 'cosmopolitan state' which are 'founded upon the recognition of the otherness of the other'.

Furthermore, the uncontrollable terrorist risks faced and subsequent deterritorialisation of risk profiles have had inevitable and dramatic effects on financial markets in general and specifically upon insurance markets with the boundaries of private insurability 'dissolving' (Beck, 2002, p.50). As Kirsch (2001) further argues 'the insurance crisis actually reflects more about something more about the risks we are living with now, including terrorism: namely, that in a society forced increasingly to share risks that are, in a sense, uninsurable'.

Beck also points to the deep social consequences of facing up to living within the new contours of the world risk society and in particular the threat of attack derived from advances in science and technology:

> But the most horrifying connection is that all the risk conflicts that are stored away as potential could now be unintentionally released. Every advance in gene technology to nanotechnology opens up a 'Pandora's box' that could be used as a terrorist toolkit. Thus the terrorist threat has made everyone into a disaster movie scriptwriter, now condemned to imagine the effects of a home-made atomic bomb assembled with the help of gene or

> nanotechnology; of the collapse of global computer networks by the introduction of groups of viruses and so on.
>
> (Beck 2002, p.46)

He continues by arguing that 9/11 has increased the perceived prospect of 'new' forms of terrorist attack, and in particular, WMD:

> There is a sinister perspective for the world after September 11th. It is that uncontrolled risk is now irredeemable and deeply engineered into all the processes that sustain life in advanced societies. Pessimism then seams to be the only rational stance. But this is a one sided and therefore truly misguided view. It ignores the new terrorism. It is dwarfed by the sheer scale of the new opportunities opened up by today's threats that is the axis of conflict in the world risk society.

Exposing the Future City

Since 9/11, the City of London has been under increased risk from terrorism with some reports even claiming that there was a detailed plan developed by al-Qaeda[1] to bomb the Square Mile (*The Observer*, 16 December 2001). For example one uncorroborated report in December 2001[2] noted that:

> Chilling plans for a devastating bomb attack on the City of London have been discovered in a terrorist base in Afghanistan, revealing a sophisticated al-Qaeda training programme to spread its campaign to Britain. The blueprint is contained in a notebook written in clear English at an al-Qaeda camp in the former Taliban stronghold of Kandahar.

The repost continues by highlighting that an area of the central City was a planned target:

> In step-by-step instructions it describes how to construct a huge remote-controlled van bomb – identical to those used by al-Qaeda against the US embassies in Kenya and Tanzania with lethal effect in 1998. A scribbled note on top of one page suggests the intended target was Moorgate in the centre of London's financial district.

Whether or not these reports are strictly accurate is a matter of conjecture. However, what is clear is that during its evolution the ring of steel strategy has undergone continual reassessment notably after the 1996 Docklands bomb and most recently after 9/11. On both occasions these events created considerable debate in policy and academic circles about the future shape and functioning of cities *per se* and in particular those cities or strategic sites within cities, faced with the imminent threat of terrorism.

Such assessments of anti-terrorist security strategies are of course not new as for many years' urban commentators have discussed the costs and benefits of adopting counter terrorism measures. For example over a decade ago Mike Davis (1992) prophesised that within the crime-infested future city the car bomb could well become the ultimate weapon of crime and terror, and predicted that the response to this potential threat would increasingly lead urban authorities and the agents of security to consider overt fortressing similar to the ring of steel around Belfast city centre in the 1970s. In more recent times acts of mega-terrorism such as 9/11 and possible attacks using WMD (such as the attack on the Tokyo underground in 1995) have led to a reassessment of counter-terrorism measures, especially with regard to the continual militarisation of the city.

Although the adopted approach in the City of London has highlighted that such a fortress approach is not necessarily required or desired, other accounts continue to highlight this method of overt militarisation as part of an overall model of future urbanism. Some commentators have argued that perhaps it was time that planners, developers and architects began to consider safety and damage limitation against terrorism when designing cities (Boal, 1975; Jarman, 1993), and in particular, in relation to individual structures that are ill-prepared to withstand a bomb blast (Haynes, 1995; Hyett, 1996; Hall, 2001).

Such accounts, in particular placed great emphasis upon 'target hardening'. For example Martin Pawley (1998, p.148) in *Terminal Architecture*, argued that as a result of an upsurge in urban terrorism, especially against 'the highly serviced and vulnerable built environment of the modern world', the new-wave of signature buildings could be replaced by an 'architecture of terror', as a result of security needs. This he argued could well have the function of making the buildings 'anonymous', and thus, he concluded, a less unattractive terrorist target. Pawley, using examples from Israel, Sri Lanka, North America, Spain and the UK, further inferred that this 'architecture of terror' will be self-reproducing as planning guidelines once drawn up will be difficult to withdraw, and such defensive architecture will become 'impossible to resist' once bombs are detonated, due to calls to reduce the impact of terrorism through urban and architectural design.

What such an account fails to take account of is that balance between security and effective business functioning, and that militarized design modifications are always desirable. As has been shown, the City of London began enhancing physical security, but only after the second major bomb in 1993. Prior to this a ring of steel-solution would have been seen as an over-reaction, and symbolic of the stigmatised Belfast-approach to containing the terrorist threat. Indeed the approach eventually adopted was far 'softer' than Pawley advocates, suggesting pragmatism on behalf of the Corporation of London with regard to balancing the need to counter terrorism with the need to promote the City as the a global financial node.

There have also been concerns raised that anti-terrorist defences, if constructed, could mean the virtual death of the urban areas as functioning entities. As architectural critic Paul Finch (1996, p.25) noted after the 1996 Docklands bomb:

> The truth is we do not design buildings to withstand bomb blasts because bomb blasts are the exception to the rule...the key point is the defining of the line between risk and recklessness. We do not generally conduct our lives on the basis of the worst thing that could in theory happen.

Furthermore, there is a need to consider the displacement of risk that the overt fortressing of certain areas could have. There has certainly been much concern about the ring of steel around the City causing a rethink of terrorist targeting and the switching to less well defended targets.

Rethinking Urbanism Post 9/11

Similar statements to those noted in the last section have been made in the aftermath of 9/11 in relation to both militarisation of financial centres and of civil society – for example through the construction of security cordons, exclusion zones, cordon sanitaire's, or ever expanding CCTV networks around strategic sites, as well as the 'surge' in the fortressing of residential environments and public spaces. However as Stephen Graham (2001, p.411) ironically points out such 'old defensive responses… seem almost comically irrelevant in this new age of threat', and in particular WMDs. Others also cautiously point out that the rise of the 'fortress city' is embedded within a pro-security discourse and power structures which require illumination. As Jennifer Light (2002, p.612) highlights, fear of terrorism and the explosion of militarised urban space 'reminds us that many powerful economic and political interests are well served by the unbridled expansion of urban fear'.

In the immediacy of the 9/11 attack the rhetoric was very much that of 'cities under siege' (Catterall, 2001), especially those strategically placed in the global economy. Mike Davis (2001) for example in an article on *The Future of Fear* argued that fear and a reduction in civil liberties will ensue, especially in American cities where 'deep anxieties about their personal safety may led millions of otherwise humane Americans to invest in the blind trust of the revamped National Security State'. Subsequently, questions are now being asked by urban governments and citizens alike about whether or not the plethora of post 9/11 security is actually significantly improving security and reducing the fear of terrorist attack? Or indeed whether we can adequately plan against similar attacks in the future?

Defensible Space Revisited – Physical and Economic Impacts of 9/11

A militarised perspective has now enveloped urban security agendas in many western cities, 'so that the inevitable fragilities and vulnerabilities they display can be significantly reduced' (Graham 2002a, p.589). As such 'military and geopolitical security now penetrate utterly into practices surrounding governance, design and planning of cities and region' (ibid.). However, 9/11 has merely signalled a surge towards an ever increasingly militarised city. As Warren (2002, p.614) notes 'it is misleading to assume that these military and paramilitary operations in urban centres began on September 11':

> Rather than a cause, the 'War on Terrorism' has served as a prism being used to conflate and further legitimize dynamics that already were militarizing urban space. These include the revision of long standing military doctrine to accept and rationalize multiple missions within the urban terrain; turning vast areas of cities into zones of video and electronic surveillance; and the repression and control of mass citizen political mobilization in cites. These phenomena have expanded and deepened in the aftermath of September 11.

As such 9/11 heralded many discussions regarding how major cities might look and function in the future if the threat of terrorism persists. In particular, there was a reassessment of the viability of building iconic skyscrapers, with some even predicting their demise altogether:

> The construction of glamorous ever-higher trophy skyscrapers will stop; the towers in Kuala Lumpur and Frankfurt have already felt the threat, closing and evaluating the day after the World Trade Center collapse; workers in the Empire State building in New York and the Sears Tower in Chicago are already reported to be afraid to go up to their offices.
> (Marcuse, 2001, p.395)

Kunstler and Salingaros (2001) in *The End of Tall Buildings* went further. They predicted that 'no new megatowers will be built, and existing ones are to be dismantled. This will lead to a radical transformation of city centers…'.

From a more rational and less reactionary perspective Mills (2002, p.200) highlights the economic arguments which might mean skyscrapers are seen as less attractive options:

> Under any positive risk of terrorist attack, tall buildings are better targets than short buildings, especially buildings that are taller than nearby buildings. Undoubtedly the insurance costs per dollar of insured value will be an increased function of building height… Profitable building height is an increasing function of land values. However, central business district (CBD) land values are high because of the value of proximity to other similar activities. Increasingly dangers of terrorist attacks reduce the value of such proximity and therefore will lead to lower CBD land values and hence to lower office building height.

Others commentators hypothesised, drawing on the work of Pawley (1998) that we might see 'the massive growth of relatively anonymous, low level fortressed business spaces' (Graham, 2001, p.414). As highlighted in previous chapters this tendency to sound the death knoll of 'building tall' has generally been proved incorrect. As Graham (2002c) further noted in relation to 9/11:

> The iconic power of the skyscraper – a symbol of urban 'progress' and modernization for a century – was instantly reversed from icon of power, progress and the dynamism of urban America, it has been transmuted into a symbol of fragility which builds deep vulnerability into he cityscape. And yet skyscraper construction continues apace and many new proposals are still emerging (p.27).

The City of London for example provides an example where there has been a significant increase in tall buildings being constructed since 9/11.

Others have also suggested that dispersal of key functions away from city centres could be one result of 9/11 and the continuing attempt to protect the economic functioning of the city. For example, Vidler (2001a) writing in the *New York Times* on 'A City Transformed: Designing Defensible Space', alluded to Oscar Newman's classic work, and hypothesised about the nature of experiencing city life:

> The terrorist attack on the World Trade Center is propelling a civic debate over whether to change the way Americans experience and ultimately build upon urban public spaces. Are a city's assets – density, concentration, monumental structures – still alluring? Will a desire for 'defensible space' radically transform the city as Americans know it?

He argued that there will be an 'understandable impulse to flee', for cities to disperse to suburbia in search of 'space and security', and that the spatial dynamics of cities would be dramatically affected (Vidler, 2001b):

> Are we looking at conditions where the need for 'defensible space' will gradually transform the form of the city as we know it? Where dispersal rather than concentration will be the pattern of life and work and where monumental forms of building will give way to camouflaged sheds, or disbursed all together into home offices? Are we about to face the collapse of the attraction of density, out of fear, the over policing of access to the public realm, or simple necessity?

Other commentators have also addressed the potential urban impacts of 9/11 highlighting both the potential desire to flee the city and the continual need for business clustering. Peter Marcuse (2002b) for example hypothesises about a number of potential consequences for urban economies and real estate industries since 9/11 which he argues will further increase the 'partitioning of urban space'. The net result he argued 'might be described as a decentralization of key business activities and their attendant services, but to very concentrated off-center locations in close proximity to the major centres' (p.596, see also Marcuse, 2000a; Marcuse and Kempton 2002). This he referred to as 'concentrated decentralisation'. He further anticipated that in both these new areas of activity and the old areas an increase 'barricading' would occur as well as a 'citadelisation' of business services and exclusive residences. In short, he notes 'security becomes the justification for measures that threaten the core of the urban social and political life, from the physical barricading of space to the social barricading of democratic activity' (Marcuse, 2000a, p.276).

Like Mike Davis's bleak portrayal of 'fortress LA' in the 1990s, Marcuse draws primarily on examples from New York post 9/11 to draw his conclusions. His general arguments do have parallels in London where after major bombings similar sentiments are common with the need to 'secure the City' high on the agenda of Government and police to stop a supposed exodus of financial institutions to more peripheral and less risky locations. Furthermore, in May 2003 an extreme example of such barricading occurred when the Houses of Parliament in London were surrounded by a series of large concrete blocks, referred to as the 'ring of concrete', to protect the building from potential car bomb attacks.

Other writers, predominantly in America, have taken a more econometric view of the potential impacts of terrorism on city life, questioning whether urban economies will continue to thrive in the face of further threats of large scale attack. As previously noted the ambiguity surrounding the withdrawal of insurance and reinsurance is highlighted as a key concern which could 'slow office construction long after the additional construction is socially justified by economic recovery' (Mills 2002; see also Glaeser and Shapiro, 2002). For example work by Harrigan and Martin (2002) noted that the cost of 9/11 for New Yorkers was astronomical both in terms of the actual disaster site but in terms of: impacts on transportation links which become slower because of increased security and bomb alerts; increased expenditure on security; massively increased insurance premiums; and also cost to productivity through emotional impact on workers fearing future attacks. As such they argued on one hand that the net impact, what they refer to as 'a terror tax' is to make cities 'less attractive'. However, on the other hand they noted that:

> The forces that lead to city formation also enable cities to be highly resilient in the face of catastrophes such as terrorist attacks, because they constitute a force for agglomeration that is very difficult to overcome (p.107).

Overall, the consensus that is now emerging suggests that the strong the economic forces that create the clustering of economic and financial functions in cities will be more than strong enough to alleviate any short term fears of terrorist attack (Swanstrom, 2002).

A Socio-technical Fix?

The perceived consequences on 9/11 for urban development focused heavily on the 'physical' changes that might occur as a result of the increased militarisation of the city. However, equally important is the potential 'social' impact'. Peter Marcuse (2001) for example argued that social polarization on the basis of income and race will be exacerbated 'with the focus of upper-income disproportionately white households concentrating in more tightly controlled citadels and others more and more excluded and segregated with sharper dividing lines between and among groups'. The same argument applies to 'public space' which he argued will become increasingly privatised and 'tightly controlled', through CCTV and regulated in terms of access restrictions. In short this means that 'democratic conduct' is an essential ingredient of a counter-terrorist response if we are to maintain the quality of city life.

Furthermore, others such as Michael Saifer (2001, p.416) have argued that a lasting response must contain the promotion of 'civic consciousness and cultural coexistence' at both local and global scales. He further agues that cities are therefore centrally placed to build a 'viable and sustainable system of 'global governance' founded on the mobilization of local and global civil society' (ibid.).

Other commentators have more explicitly argued that technological advancement will become all-important in the battle against urban fear and terrorism. There is already widespread evidence that the 'creep' of surveillance and other methods of social control in western cities in response to security concerns is begin to 'surge' in response to the new terrorist threat (Wood 2002). This is similar to the rapid expansion of surveillance technologies in the UK in the early 1990s after a spate of terrorist bombings and other criminal activities. In the contemporary city there are fears that the mushrooming of increasingly automated and hi-tech systems will further erode civil liberties as democratic and ethical accountability will be given a back seat in the new era of 'anxious urbanism' (Farish, 2002) which has followed 9/11. In short, post 9/11 there has been a commodification of surveillance with a technological drive to develop digital, automated and biometric systems (Lyon, 2002a, 2002b).

In the city, surveillance technologies are increasingly leading to the automatic production of space with urban society quickly becoming a technologically managed system based on automated access and boundary control. One of the most influential pieces of work in this area is the idea of Automated Socio-Technical Environments (ASTEs), developed by Lianos and Douglass (2000). They argued that such environments occur as a result of the pervasive 'dangerisation' of society and the generation of environments of risk. ASTEs are seen as high-tech risk management devises with a number of key features. First, the user cannot negotiated with the system as the system is fully automated involving no human interaction. Second, access to the system is accomplished through specific legitimising principles. As such:

> For the system there are no good and bad, honest and dishonest – or for that matter, poor and less poor individuals. There are simply holders of valid tokens for each predetermined level of access.
>
> (ibid. p.265)

It is argued that ASTE have the potential to radically change the social and technological infrastructure of cities as they become reconfigured according to ever-changing management priorities, distinguishing and discriminating only on the ground of 'quality of user' rather than other social categories such as age, race and sex which the system does not recognised – 'you either insert the right ticket or you do not' (ibid. p.266). Graham (2002, p.241) for example, highlights that such systems could 'start to inscribe normative ecologies of acceptable people and behaviour' within the city. Employing the metaphor of 'redlining' Graham further notes that this could well mean that:

> ...individuals will be excluded from venturing into the premium commercial spaces of their city due to their appearance, habits or challenges to the dominant power holders' normative concepts of who belongs where and when within the city.
>
> (ibid.)

ASTEs are in short both 'a mode of social control and management tool where suspicion is simply pre-embedded in the system' (Lianos and Douglass, 2000, p.269):

> Automated regulatory procedures are not simply procedures of control. They are general management instruments for adapting the social world to the aims of the institutions that use them. There purpose is to eliminate all those aspects of social information which prevent the institution from achieving its set targets. This is why automated environment operate on the basis of suspected potential dangers caused by their users (p.266).

As technology develops apace it is likely that urban areas will see ever increasing number of ASTEs with security concerns being integrated into the automated environment. Perhaps most notably this will occur with CCTV technology:

> The diffusion and automation of CCTV, and its linkages to digital databases, however, means that the normative assumptions about the value and risk associated with particular individual's moves from the discretion of human practice to be embedded within the opaque codes of computer systems.
>
> (Graham, 2002, p.240)

From the point of view of this book, the Automated Number Plate Recognition systems used by the City of London Police since 1997 provide a perfect example of such ASTEs, which according to some (Rosen, 2001) will be reinforced by biometric (facial recognition) cameras in the near future. Furthermore attempt to reduce traffic gridlock in central London in 2003 by setting up a 'congestion change' zone has further extended the geographical boundaries of the City's ASTE which is ringed with similar automated 'charging' devices with double-up as counter-terrorism systems (Coaffee, 2002). These systems are certainly ones that the agencies of security, most notably the police, as well as insurers will endorse fully as it promotes safety, retains movement flows and meets with commercial imperatives (Lianos and Douglass, 2000, p.270).

Further accounts of the potential role of ASTE's take such systems to other technologically deterministic and dystopian extremes. Huber and Mills (2002) in an article for *City Journal*, entitled *How Technology Will Defeat Terrorism*? gave some applications of ASTE's in the post 9/11 city. They highlight that 'step by step, cities like New York must learn to watch and track everything that moves' and to have the ability to recognise threats through 'a massive deployment of digital technology' which can screen anything from vehicles, people, letters and certain types of smell and 'dust' particulates. They note that today the technology is now available and cost effective:

> A decade ago, none of this would have been economically feasible. It is today. It will entail a lot of new investment; but the technology is there, or very close to there – real, commercial, functional. It's going to get deployed, not only at airports but even more widely on private premises and, later, for municipal use.

They also highlight that civil libertarians will complain about invasion of privacy and restriction of access but argue that this is a reality that we have been living with for years with toll roads, paying to get in to civic buildings such as sports stadium and screening people as they go into corporate buildings such as museums and schools. Finally they argue emotively that micro scale surveillance is the key to terrorist threat recognition:

> Properly deployed at home, as they can be, these technologies of freedom will guarantee the physical security on which all our civil liberties ultimately depend. Properly deployed abroad, they will destroy privacy everywhere we need to destroy it.

They continue:

> It may seem anomalous to point to micro-scale technology as the answer to terrorists who brought down New York's tallest skyscrapers. But this is the technology that perfectly matches the enemy's character and strength. It can be replicated at very little cost; it is cheap and expendable. Small and highly mobile, it can be scattered far and wide... It is a horrible vision. It gives us no joy to articulate it. But at home and abroad, it will end up as their sons against our silicon. Our silicon will win.

Contemplating Urbicide

The preceding sections have highlighted the potential physical and economic and, social and technological impacts of new forms of terrorist risk. The approaches outlined highlight the way in which cities affected or at risk from terrorist attack can *re*construct and *re*generate themselves. Some of these approaches adopted are little more than symbolic gestures with little power of enforcement (for example signs saying you are entering a security cordon) or the development of some form of enhanced civic cosmopolitanism. Others are based on managerial techniques of control (such as stop and search procedures). Others are more overt and directly restrict access to a territory (for example security gates) or keeping the occupants of an area under constant surveillance (such as automated CCTV) or in extreme cases lead to the relocation of business to the city limits and beyond. Such approaches are increasingly using military technologies to improve both the real and perceived security of cities.

However, taken to an extreme the militarisation of urban space as a result of attempts to counter terrorism can lead to the *de*struction and *de*generation of urban centres in what can be referred to as 'urbicide' – 'the deliberate wrecking or killing of the city' (Graham, 2003, p.63; see also Berman, 1996; Saifer, 2001; Coward, 2002).[3] In relation to terrorism the concept of urbicide is double edged. It can be seen as a strategy of the terrorist, for example the targeting of cities as strategic sites in order to destroy its physical infrastructure, damage its economy and perpetuation of fear throughout the urban population. As Saifer (2001) noted the aim of such insurgency strategies is 'to destroy the security, public order, civility and quality of life of all their citizens, and damage or destroy the viability and liveability of the city itself' (p.422). On the other hand, urbicide has been shown to have been used as a pre-emptive counter terrorist strategy to demolish, and literally annihilate urban spaces in order to destroy the networks and infrastructure of terrorist organisation residing in cities. For example, Stephen Graham (2003) in an paper entitled *Lessons in Urbicide*, argued that such destruction was the explicit strategy of the attempts by the Israeli Defence forces (codenamed 'operation defensive shield') to systematically bulldoze a large area of a refugee camp in Jenin in the occupied territories of the West Bank and was an attempt to 'destroy the urban foundations of the proto-Palestinian state' (p.63). He further argued that this was a 'deliberate and systematic destruction', and forced 'de-modernisation' of key areas of urban space. In this sense he argued that 'Palestinian cities are portrayed as potentially impenetrable, unknowing space, which challenge the three-dimensional gaze of...

high technology surveillance systems and lie beyond much of its heavy duty weaponry' (pp.70-1). Extrapolating this idea to the wider 'war on terrorism' Graham argues that scenes such as those in Jenin are likely to be repeated in the future as 'the enemy', faced with the increasing technological capabilities of modern weaponry to detect and destroy, will seek sanctuary in cities where they can more easily camouflage themselves.

> The world is becoming patchwork of cities. Increasingly, there is nowhere else to go. In such a setting, cities are indeed the natural habitat of anyone who falls within the ever-widening search for targets and opponents in the war on terror (p.77).

Concluding Comments on the City of London

Urban researchers are increasingly aware that processes which occur in one city are locally specific, and should not be mapped uncritically onto other areas. Great stress is now placed on the multiple discourse communities that exist within cities, as well as upon local circumstances of place, noting that general urban transformations are occurring unevenly in different cities. The case of the City of London has been theoretically situated in this book to avoid such normative assumptions. It has argued that the strategies by which the City adapted to the risk of terrorism were contingent upon local histories and geographies, and in particular by a combination of time dependent processes acting within, and upon, this defined geographical area. These processes provided the 'frames of reference', the specific socio-economic, political and institutional context, in which the key urban managers operated, and which served to enable and constrain their actions and strategies.

As such the adaptations made to the urban landscape in the City as a response to terrorist risk were a result of a unique sets of local circumstances related to the Square Mile which set it apart from other areas, apart from than at a general level of abstraction. In particular this relates to: the specific nature of its institutional thickness; the influence of the private sector in its affairs; the City's tremendous power and influence in the governance of London; the high concentration of valuable buildings; the City's compact urban morphology; and the increasingly cosmopolitan business community in the Square Mile.

Whilst recognising the inherent difficulties of transplanting the conclusions from this research into broader urban debates, a number of findings from this work illuminate the issues around a number of defensive urban landscapes and urban processes: approaches to 'hardening' landscapes; the relationship between landscape form and fear; the influence of new institutional mechanisms and strategic collaborations in urban planning and policy decision-making; how certain landscapes are seen as global reference points; how urban form can be seen as a negotiated statement between powerful urban actors and their differing agendas; how defensive measures and insurance policy are increasingly important regulators of the urban landscape; and how defended enclaves increasingly contribute to the spatial restructuring of the city. These are debates which have been ongoing through the 1990s and have been especially prevalent after 9/11.

The City's initial response to terrorism in the 1990s was very much related to the time-period of the attacks during the early 1990s when attacks against prime economic targets, especially in economically advanced Western cities, were increasing. In particular, in 1988 there was a clear shift in Provisional IRA strategy towards attacking 'economic' targets in England. The detonation of large bombs in the City in 1992 and 1993, as well as a number of other attempts against this area, demonstrate the Provisional IRA's recognition of the important place of the City in the national and global economy.

In the last decade a number of attempts to design out terrorism have been introduced in the Square Mile including physical barriers to restrict access, advanced electronic surveillance in the form of three interrelated cameras networks and insurance regulations and blast protection measures as well as innumerable measures that were operationalised by activating individual and community responses. In addition, as a result of the terrorist threat, new governance arrangements have been developed within the City, as well as between the Corporation of London, surrounding local authorities and Central Government. The establishments of these new 'institutional pillars' necessitated by the terrorist risk served to reinforce the City's economic position at a time when the leaders of the Square Mile were being criticised for failing to keep pace with the rapid changes taking place in the global economy.

Through the development of a series of defensive strategies the City put itself in a better position to prevent another major bombing incident. The security arrangements were judged a success, at least at a corporate level, not only because they have, to date, prevented a third major bomb which would have significantly effected the City's international reputation, but ironically also because of the positive by-products of the scheme which saw crime, atmospheric pollution and traffic congestion and accidents decrease, and major improvements made in business disaster recovery planning. However, the identity of the City has been irreparably changed by terrorist attack and as a consequence of its counter-response. It became seen as a place where the demands of the global finance industry have been merged with a culture of security to create a successful and safe working environment.

The anti-terrorist security strategies employed in the City of London have been subject to both praise and criticism. Michael Cassidy, the chief architect of the City's effort to 'beat the bombers', summed up the ring of steel in the following way noting the cordon's advantages both for business confidence and the enhancement of environmental quality:[4]

> Totally positive in all areas. It's given people an extra comfort factor. The trend of banks coming here was not interrupted at all by the last two bombs, in fact the strength of the trend has built up ever since...None left for security reasons. The typical City worker finds that the environment is better and from that point of view they wouldn't want to see it go. Not now.

This however only tells part of the story. It fails to recognise the associated effects of traffic and risk to bordering areas, the concerns of civil liberty groups about the privatisation of public space, and the paternalistic way in which the scheme was constructed without what many felt was adequate consultation.

This enquiry has demonstrated that attempts to reduce the risk of terrorism in the City of London were undertaken by a variety of different groupings and stakeholder coalitions, most notably the City of London police, the insurance industry and the Corporation of London, as well as private businesses interests. The strategies adopted by these groupings can therefore be read as an inscription of economic and political power symbolising the City as an economic terrorist target, a defended enclave as well as an adaptive space of global finance.

As a result of terrorist attack, and the risk of further bombings in recent years, a counter-terrorist strategy to control and regulate the space within the Square Mile has been adopted and refined which attempts to balance security with business continuity. This is highlighted by one of the City of London Police's local priorities in relation to their terrorism strategy:

> The City of London policing style is aimed at the prevention and detection of terrorist attacks with the use of improved technology and high profile patrols. Our target is to maintain the current level of vigilance and thereby deter and future terrorist criminality. The Terrorism Strategy will achieve these aims through high visibility policing, directed intelligence, technology and partnership.
>
> (City of London Police, 2002c)

In particular the City of London Police's 2002-7 Corporate Plan (City of London Police, 2002d) highlighted six areas for countering terrorist activity:

- Targeting police officer deployment focusing on high visibility uniform patrols within the city in particular with officer utilising extended powers under the Terrorism Act 2000;
- Effective use of available and emerging technology in respect to CCTV and automatic number plate recognition (including mobile units) to support our operational policing tactics and to aid post incident investigation;
- To utilise all available intelligence for the implementation of intelligence-led counter terrorist tactics for threat assessment and pro-active operations;
- To work in partnership with the community to achieve our strategic aim. This includes working with formal bodies such as the Greater London Authority as well as Crime Prevention Associations and the continual rolling out of schemes such as Pager Alert and E-mail alert;
- To work with other police organisations (Metropolitan Police, British Transport Police and ministry of Defence Police) and recognised emergency services within London to ensure a co-ordinated approach to deterring criminal terrorist activity across the capital;
- To work with other agencies (such as the London Emergency Services Liaison Panel) to deter criminal terrorist activity and to provide appropriate information and guidance to the city community and that of the capital in general.

The areas noted above are similar to the rolling programme of measures which began with the bombing in April 1992 and encompassed the introduction of a number of armed road checkpoints, the imposition of parking restrictions, the fortification of individual buildings and three interrelated camera networks. The City now has well over 1500 surveillance cameras operating in the Square Mile most notably 52 high resolution Automatic Number Plate Recognition cameras which are situated strategically around the area (City of London Police, 2000a, 2002b).

The threat of terrorism that the City has to face since the early 1990s increasingly sought to focus attention on how the Corporation has adapted to modern conditions, and maintained its influence within global finance. The ring of steel, or 'ring of confidence' helped the City to create a secure platform upon which it could continue to develop and adapt its role as the financial heart of a 'world city'. The City embraces inclusion in the globalisation process whilst at the same time excluding themselves from the rest of central London through their territorial boundedness and fortification strategies. The situation that developed in the City can therefore be seen in terms of a condition of global connection and local disconnection - a condition, which, as a result of terrorism, continued to characterise the dislocated nature of the City's relationship with the rest of London both physically (Coaffee, 1996b, 2000b, 2002; Graham and Marvin, 2001) and technologically through surveillance (Coaffee, 1996a, 2002; Power, 2001, Norris and Armstrong, 1999).

For the majority of occupiers in the City the counter-terrorist strategy has been viewed positively not only as an effective security approach but also as an effective crime prevention initiative, an efficient traffic management measure and a beneficial environmental policy (Coaffee, 2000a). Statistically, the ring of steel appears to be more about crime prevention. For example Rosen (2002) highlights that the CCTV operation in the City has never caught any terrorist and in fact 'spend most of their time following car thieves and traffic offenders'.[5] Figures for 2001/2 highlight that over 12000 'offences' were detected using this CCTV system with over 6000 prosecutions made.[6]

Perhaps of greater immediacy, we should consider the surveillance implications of 9/11 and in particular a future that is likely to be pervaded by biometric camera Rosen (2001) noted that the City of London is thinking about using a biometric database for face recognition which would not only include terrorists but also 'all British citizens whose faces are registered with the national driving license bureau' (p.6). This has severe civil liberty implications. In short, 'biometrics is a feel-good technology that is being marketed based on a false promise – that the database will be limited to suspected terrorist (ibid. p.7).

These wider applications of the ring of steel means that the measures in place will remain a concrete part of the contemporary urban landscape in the Square Mile for the foreseeable future. Just as importantly from a public and social policy viewpoint, the ring of steel is now seen by business coalitions, motoring organisation, commuters, residents, neighbouring local authorities as part of London's daily life.

The events of 9/11 once again have focused attention on how the City of London, both as an independent entity and as an integral part of a global city, is responding to new terrorist threats. In recent years the Square Mile, through the use and updating of pre-existing protective security strategies have sort to reappraise their approach to counter terrorism. What has emerged is a security set up very similar to pre-9/11 security given that conventional security controls would have little impact against new terrorist realities such as WMDs. As such, in early September 2003 the City of London was 'closed down' in an elaborate anti-terrorism exercise as a response to continual threats of attack, and in particular, the second anniversary of 9/11. In this exercise over 500 of London Emergency Services personnel were tested for their state of preparedness and ability to use appropriate equipment in the event of a chemical or biological attack. This exercise focused on a hypothetical release of chemical agents on the London underground around the Bank tube station in the heart of the City of London. This test was in many ways using the Tokyo 1995 scenario when twelve people were killed and many injured by the release of the nerve agent sarin into the central Tokyo subway.[7]

Furthermore, a research report published in April 2003 highlighted that terrorism is still perceived as a significant threat by those who work in the City with almost one in ten considering the threat of terrorism on a daily basis.[8] In the City much attention is given by the Police to maintain the culture of security and vigilance.

> Total - 100% - security can never be guaranteed. But vigilance on the part of the public is one of the best ways of preventing and detecting crime, including terrorist crime. We firmly believe that communities defeat terrorism. Please let us know if you have any suspicions. We don't mind if it proves to be a false alarm!
>
> (City of London Police, 2003)

This echoes national UK guidelines and rhetoric which seeks to provide a balance between democracy and appropriate risk and security management responses (*Counter-Terrorist Action since September 2001* - Report to Parliament, 9 September 2002):

> We are mindful of the desire and the need of people in a vibrant democracy like ours to live normal lives without a sense of constant fear. We also know that in part because the terrorist want us to live in fear, and want to damage our economy, and the well being of our people, that they are capable of feeding false information to us in the hope that we over react...
>
> Getting the balance right is not easy...It is a task government cannot face alone. We face it with our allies and partners...and we must face it with the people of this country so that we have some shared understanding of these new, more complex terrorist threats, and a shared commitment to facing up to them: through vigilance, through support for the security authorities, and through an understanding of the difficult decisions that have to be faced by government.

Ultimately, as Swanstrom (2002) noted, 'the main threat to cities comes not from terrorism but from the policy responses to terrorism that could undermine the freedom of thought and movement that are the lifeblood of cities' (p.135). Responding to many calls for decentralization of urban functions in the post 9/11 era, he continues:

> We should resist calls to further spread out our metropolitan areas in the wake of terrorist attack. Although large and dense cities offer tempting targets for terrorists, they are also the source of our great strengths – economically, socially and politically. To abandon our cities would be to play into the hands of the terrorists (p.136).

Notes

1 The terrorist group/network believed to be responsible for the 9/11 attacks.
2 See *The Times*, 'Bin Laden in plot to bomb London City', 24 December 2001. See also *The Observer*, 16 December 2001.
3 This term was first used by a group of Bosnian architects to refer to the destruction of Balkan cities in 1992 and then by Marshal Berman in 1996 to highlight the destruction of the New York Bronx.
4 Interview conducted by the author in 1996.
5 By 1998 340 arrests and 359 stolen vehicles had been triggered by the ANPR CCTV system, all non-terrorist related (Graham and Marvin, 2002).
6 Furthermore The PACE Act led to over 3500 person and vehicle searches and over 1000 arrests (City of London Police, 2002a). In addition in August 2002 it was announced that in the preceding 18 months City Police officers had stopped more than 1800 people under the anti-terrorists legislation compared with 1056 in the previous period. However only 52 arrests were made, up from 13.
7 This test was due to take place earlier in 2003 but the second Gulf war meant it was postponed.
8 See *Facing the Future in the City* (Chiumento consultancy), 30 April 2003.

Bibliography

Adam, B., Beck, U. and van Loon, J. (2000), *The Risk Society and Beyond*, Sage, London.

Adams, J. (1995), Risk, UCL Press, London.

Advisory Committee to the NAIC Redlining Task Force (1978), Ninety Day Report to the Advisory Committee to the NAIC Redlining Task Force, National Association of Insurance Commissioners, Milwaukee.

Agnew, J. A. (1987), *Place and politics: the geographical mediation of state and society*, Allen and Unwin, Boston.

Agnew, J. and Corbridge, S. (1995), *Mastering Space - hegemony, territory and international political economy*, Routledge, London.

Alderson, A. Adams, J. and Clarke, L. (1993), 'The carnage we cannot control', *Sunday Times*, 2 May, p. 12.

Alexander, Y. and Latter, R. (eds.) (1990), *Terrorism and the media – dilemmas for government, journalists and the public*, Brassey's, London.

Amin, A., and Graham, S. (1999), 'Cities of connection and disconnection', in Allen. J. Massey, D. and Pryke, M. *et al*, *Unsettling Cities*, Routledge, London.

Amin, A. and Thrift, N. (eds.) (1994), *Globalization, institutions and regional development in Europe*, Oxford University Press, Oxford.

Amin, A and Thrift, N. (1995), 'Living in the global', in Amin, A. and Thrift, N. (eds.) *Globalization, Institutions and Regional Development in Europe*, Oxford University Press, Oxford, pp. 1-22.

Anderson, J. (1996), 'The shifting stage of politics: new medieval and postmodern territorialities?', *Environment and Planning D: Society and Space*, Vol. 14, pp. 133-53.

Anderson, J. and Shuttleworth, I. (1998), 'Sectarian demography, territoriality and political development in Northern Ireland', *Political Geography*, Vol. 17(2), pp. 187-208.

Anderson, P. (1987), 'The figures of decent', *New Left Review*, Vol. 161, pp. 20-77.

Appadurai, A. (1990), 'Disjuncture and difference in the global cultural economy', *Theory, Culture and Society*, Vol. 7, pp. 295-310.

Applied Property Research (1993), *Current security arrangements in the City*.

Archibald, R., Jamison, J., Rosen, B. and Schachter J. (2002), *Security and Safety in Los Angeles High-rise Buildings after 9/11*, Rand, Santa Monica.

Ardrey, R. (1966), *The territorial imperative*, Antheum, New York.

Ardrey, R. (1970), *The social contract*, Delta, New York.

Ashworth, J. (1996), 'Planner who built his name on controversy', *The Times* (Internet edition) 31 December.

Association of London Authorities (1994), *At a premium – the cost of home contents insurance to Londoners*, March.

Atkins, P., Simmons, I. and Roberts, B. (1998), *People, land and time*, Arnold, London.

Atkins, P.J. (1993), 'How the West End was won: the struggle to remove street barriers in Victorian England', *Journal of Historical Geography*, Vol. 19(3), pp. 265-77.

Ayers, B. (1994), 'The Peacemaker Returns', *Citywatch – The City of London Police Magazine*, Winter, p. 3.

Badain, D. (1980), 'Insurance redlining and the future of the urban core', *Columbia Journal of Law and Social Problems*, Vol. 16, No. 1, pp. 1-85.

Badcock, B. (1996), 'Looking glass views of the city', *Progress in Human Geography*, Vol. 20(1), pp. 91-9.

Baerwald, T. (1981), 'The site selection process of suburban residential builders', *Urban Geography*, Vol. 2, pp. 339-57.

Bagnall, S. (1993), 'Companies may face compulsory terrorism levy', *The Times*, 27 April, p. 24.

Bailey, N. (1999), 'Putting local communities at the heart of regeneration: the rhetoric and the reality', paper to a seminar on *Institutional capacity, social capital and urban governance*, Newcastle upon Tyne, April.

Baily, E. (1993), 'No desks or buildings – but business as usual', *Daily Telegraph*, 26 April, p. 3.

Ball, M. (1998), 'Institutions in British Property Research: A Review', *Urban Studies*, Vol. 35(9), pp. 1501-17.

Banham, R. (1973), *Los Angeles: The Architecture of Four Ecologies*, Penguin, London.

Bannister, J. (1991), 'The London Insurance Market and its Future', *The Journal of the Chartered Insurance Institute*, March, pp. 8-12.

Barnes, T. and Duncan J. (eds.) (1991), *Writing Worlds: discourse, text and metaphor in the representation of landscape*, Routledge, London.

Barr, R. and Pease, K. (1990), 'Crime placement, displacement and deflection', in Tonry, M. and Morris, N. (eds.), *Crime and Justice: a review of research*, University of Chicago, Chicago, Vol. 12, pp. 227-318.

Barton, B (1996), *A Pocket History of Ulster*, O'Brian Press, Dublin.

Batley, R. (1980), 'The political significance of administrative allocation', in *Papers in Urban and Regional Studies 4*, CURS, Birmingham, England, 17-44.

Bauman, Z. (2002), 'Reconnaissance Wars of the Planetary Frontierland', *Theory, Culture and Society*, Vol. 19(4), pp. 81-90.

Beauregard, R.A. (1999), 'Break Dancing on the Santa Monica Boulevard', *Urban Geography*, Vol. 20(5), pp. 396-9.

Beazley, M., Loftman, P. and Nevin, B (1997), 'Downtown Development and Community Resistance: an international perspective', in Jewson, N. and MacGregor, S. (eds.) *Transforming Cities: contested governance and new spatial division*, Routledge, London, pp. 181-92.

Beck, U. (1992a), *Risk Society - towards a new modernity*, Sage, London.

Beck, U. (1992b), 'From industrial society to the risk society: questions of survival, social structure and ecological enlightenment', *Theory, Culture and Society*, Vol. 9, pp. 97-123

Beck, U. (1995), *Ecological politics in an age of risk*, Polity Press, Cambridge.

Beck, U. (1997b), 'The relations of definitions: The cultural and legal contexts of media constructions of risk', paper presented at *The Media, Risk and Environment Symposium*, University of Wales, Cardiff, 3-4 July.

Beck, U. (1999), *World Risk Society*, Polity Press, Cambridge.

Beck, U. (2000), 'The cosmopolitan perspective: sociology of the second age of modernity', *British Journal of Sociology*, Vol. 51(1), pp. 79-105.

Beck, U. (2001), 'The Cosmopolitan state', *Eurozine*, 12 May.

Beck, U. (2002), 'The Terrorist Threat: World risk society revisited', *Theory, Culture and Society*, Vol. 19(4), pp. 39-55.

Beck, U. Giddens, A. and Lash. S (1994), *Reflexive Modernisation: Politics, Tradition and Aesthetics in the Modern Social Order*, Polity Press, Cambridge.

Belussi, F. (1996), 'Local system, industrial districts and institutional networks: towards a new evolutionary paradigm of industrial economics?', *European Planning Studies* Vol. 4(3), pp. 5-26.

Benton, L. (1995), 'Will the real/reel Los Angeles please stand up?', *Urban Geography*, Vol. 16(2), pp. 144-64.

Berger, J. (1972), *Ways of Seeing*, BBC Publications, London.

Berman, M. (1996), 'Falling Towers: City life after urbicide', in Crow, D. (ed.), Geography and Identity, Maisoneuve Press, Washington, pp. 172-92.

Best, J. and Luckenbill, D. (1980), 'The social organisation of deviants', *Social Problems*, Vol. 28, pp. 14-31.

Best, S. and Kellner, D. (1991), *Postmodern Theory: Critical Interrogations*, Guilford Press, New York.

Bice, W. B. (1994), 'British Government reinsurance and acts of terrorism: the problems with Pool Re', *University of Pennsylvania Journal of International Business Law*, 15(3), pp. 441-68.

Binney, M. (1996), 'These beats are made for walking', *The Times* (Internet edition), 13 March.

Bishop, P. and Mallie, E. (1987), *The Provisional IRA*, Corgi, London.

Blackhurst, C., Thompson, R. and Warner, J. (1993), 'City's reputation around the world is put at risk', *Independent on Sunday*, 25 April, p. 3.

Blakely, E.J. and Snyder M.G. (1995), *Fortress America: gated and walled communities in the US*, Lincoln Institute of Land Policy, Working Papers.

Blakely, E.J. and Synder, M.G. (1999), *Fortress America: Gated Communities in the United States*, The Brookings Institution, Washington.

Blowers, A. (1997), 'Environmental policy: ecological modernisation or risk society?', *Urban Studies*, Vol. 34, pp. 845-71.

Blowers, A. (1998), 'Nuclear waste and landscapes of risk', Keynote address at the *Landscapes of Defence Conference*, Oxford Brookes University, 31 May.

Blowers, A. (1999), 'Nuclear waste and landscapes of risk', *Landscape Research* 24(3) pp. 241-64.

Boal, F.W. (1969), 'Territoriality on the Shankhill-Falls Divide, Belfast', *Irish Geography*, 6, pp. 30-50.

Boal, F.W. (1971), 'Territoriality and class: a study of two residential areas in Belfast', *Irish Geography*, 6, pp. 229-48.

Boal, F.W. (1975), 'Belfast 1980: a segregated city?', *Graticule,* Department of Geography, Queens University of Belfast.

Boal, F.W. (1995), *Shaping a City - Belfast in the late Twentieth Century*, Institute of Irish Studies, Belfast.

Boal, F.W. (1996), 'Integration and division: sharing and segregation in Belfast', Oxford University, *Department of Geography Research Seminar*, 2 February.

Boal, F.W. and Douglas, J.N.H. (eds.) (1982), *Integration and Division - Geographical Perspectives on the Northern Ireland Problem*, Academic Press, London.

Boal, F.W. and Murray, R.C. (1977), 'A City in Conflict', *Geographical Magazine*, Vol. 49, pp. 364-71.

Bonshek, J. (1990), 'Why the skyscraper?: towards a theoretical framework for analysis', *Environment and Planning B: Planning and Design*, 17, pp. 131-48.

Bosovic, M. (ed.) (1995), *The Panopticon Writings – Introduction*, Verso, London.

Bottoms, A.E. (1974), 'Review of defensible space', *British Journal of Criminology*, 14, pp. 203-6.

Bowers, S. (2002), 'Minerva's towering ambition', *Guardian Online*, 21 June.

Box, S. (1988), 'Explaining fear of crime', *British Journal of Criminology*, Vol. 28(3), pp. 340-56.

Boyd, W. (1993), Book review of 'Risk Society', *Economic Geography*, Vol. 69, pp. 432-6.

Boyle, M. and Hughes, G. (1995), 'The politics of urban entrepreneurialism in Glasgow', *Geoforum*, Vol. 25(4), pp. 453-70.
Brackett, D.W. (1996), *Holy Terror: Armageddon in Tokyo*, Weatherhall, New York.
Bradbury, M. (1995), 'From here to modernity', *Prospect*, December, pp. 34-9.
Brannen, J. (1988), 'The study of sensitive subjects', *Sociological Review*, 36, pp. 552-63.
Brantingham, P. and Brantingham, P. (1984), *Patterns in crime*, Macmillan, New York.
Brantingham, P. and Brantingham, P. (1990), 'Situational crime prevention in practice', *Canadian Journal of Criminology*, 32, pp. 17-40.
Brenner, N. (1999), 'Globalisation as reterritorialisation: the re-scaling of urban governance in the European Union', *Urban Studies*, Vol. 36(3), pp. 431-51.
Brewer, J.D. (1990a), 'Sensitivity as a problem in field research: a study of routine policing in Northern Ireland', *American Behavioural Scientist*, 33, pp. 578-93.
Brewer, J.D. (1990b), *Inside the RUC: routine policing in a divided society*, Oxford University Press, Oxford.
Brewer, J.D. (1993), 'Sensitivity as a problem in field research: a study of routine policing in Northern Ireland', in Renzetti, C.M. and Lee, R.M. (eds.) *Researching sensitive topics*, Sage, London, pp. 125-46.
Brosseau, M. (1995), 'The city in textual form: Manhattan transfer's New York', *Ecumene*, Vol. 2 (1), pp. 89-114.
Brown, B. (1995), 'CCTV in town centres: three case studies', *Crime Detection and Prevention Series Paper 68*, HMSO, London.
Brown, P. (1990), 'Disney Deco', *New York Times Magazine*, 8 April, pp. 18-24, pp. 42-3, p. 48.
Brown, P.L. (1995), 'Designs in a land of bombs and guns', *New York Times*, 28 May.
Brown, S. (1984), *Retail Location and Retail Change in Belfast City Centre*, unpublished PhD thesis, Queens University, Belfast.
Brown, S. (1985a), 'City centre commercial revitalisation: the Belfast experience', *The Planner*, June, pp. 9-12.
Brown, S. (1985b), 'Central Belfast's security segment - an urban phenomenon', *Area,* Vol. 17(1), 1-8.
Brown, S. (1987), 'Shopping centre development in Belfast', *Land Development Studies*, 4, pp. 193-207.
Bruce, S. (1992), 'The problems of pro-state terrorism: Loyalist paramilitaries in Northern Ireland', *Terrorism and Political Violence*, 4, pp. 67-88.
Buckingham, C. (1994), *Terrorism: the financial implications for occupiers in the City of London*, unpublished undergraduate dissertation, Oxford Brookes University.
Budd, L. (1995), 'Globalisation, territory and strategic alliances in different financial centres', *Urban Studies*, Vol. 32(2), pp. 345-60.
Budd, L. (1998), 'Territorial competition and globalisation: Scylla and Charybdis of European Cities', *Urban Studies*, Vol. 35(4), pp. 663-85.
Burgess, E.W. (1925), The growth of the city', in Park, R.E, Burgess, E.W. and McKenzie (eds.) *The City: suggestions of investigation of human behaviour in the urban* environment, University of Chicago Press, Chicago, pp. 47-62.
Burgess, J. (1992), 'The art of interviewing', in Rogers, A., Viles, H. and Goudie, A, *The Student's Companion to Geography*, Blackwell, Oxford, pp. 207-13.
Burgess, J. (1999), 'Environmental management and sustainability', in Cloke, P., Crang, P. and Goodwin, M. (eds.) *Introducing Human Geographies*, Arnold, London, pp. 141-150.
Burns, J. (1993), 'IRA exploited reduction in spot security checks', *Financial Times*, 26 April, p. 7.
Burns, J., Atkins, R., Gapper, J., and Waters, R. (1993), 'Police seek wider power to combat terrorism', *Financial Times*, 27 April, p. 1.

Burton, I., Kates, R.W. and White, G.F. (1978), *The environment as hazard*, Oxford University Press, Oxford.

Burton, I., Kates, R.W. and White, G.F. (1993), *The environment as hazard* (Second Edition), The Guilford Press, London.

Byson, J. and Crosby, B. (1992), *Leadership in the common good*, Jossey Bass, San Francisco.

Cadman, D. and Catalano, A. (1983), *Property development in the UK - evolution and change*, College of Estate Management, Reading.

Caldeira, T.P.R. (1996), 'Building up walls: the new pattern of spatial segregation in Sao Paulo', *International Journal of Social Science*, Vol. 48(1), pp. 55-66.

Campbell, B. (1992), *Goliath: Britain's dangerous places*, Methuen, London.

Carolan, M. (1992), 'Prince hints at move over bomb cover for buildings', *Irish News*, 16 December.

Carter, S. (1993), 'Risk and the new modernity', *Postmodern Culture*, Vol. 3(3), May (Internet edition).

Cassidy, M. (1993), 'Try asking for answers to City bomb', *Estate Times*, 11 June.

Cassis, Y. (1985a), 'Bankers in English society in the late nineteenth century', *The Economic History Review*, XXXVIII, pp. 209-229.

Cassis, Y. (1985b), 'Management and strategy in the English joint stock banks 1890-1914', *Business History*, XXVII, pp. 301-315.

Cathcart, B. (1993), 'Running the Mile', *Independent on Sunday*, 2 May, p. 19.

Catterall, B. (2001), 'Cities Under Siege: September 11th and after', *City*, Vol. 5(3), p. 383.

Catterall, B. (2002), 'It all came together in New York?: Urban Studies and The Present Crisis', *City*, Vol. 6(1), pp. 145-155.

Cave, A. (2002) 'Terror takes its toll on Lloyd's', *Electronic Telegraph*, 11 April 2002.

Cawson, A. (ed.) (1986), *Organised interests and the state*, Sage, London.

Chalk, P. (1996), *West European terrorism and counter-terrorism: the evolving* dynamic, Macmillan, Basinstoke.

Chartered Institute of Transport (1991), *London transport: the way ahead*, June.

Chermayeff, S. and Alexander, C. (1966), *Community and privacy - towards a new architecture of humanism*, Penguin, Middlesex.

Cheshire, P.C. and Gordon, I. R. (1993), *European integration: territorial competition theory and practice*, Working Paper No.2, Centre for the Study of Advanced European Regions, Faculty of Urban and Regional Studies, University of Reading.

Christopherson, S. (1994), 'The fortress city: privatized spaces, consumer citizenship', in Amin, A. (ed.) *Post-Fordism - a reader*, Blackwell, Oxford, pp. 409-27.

Cho, M. (1996), 'An empirical analysis of property appraisal and mortgage redlining', *Journal of Real Estate Finance and Economics*, Vol. 13(1), pp. 45-55.

Cisneros, H. (1995), 'Defensible space: reducing crime by design', *Affordable Housing Issues*, Vol. VI(3), April.

City of London Police (2002a), *Annual Report for 2001/02*.

City of London Police (2002b), *Crime and Disorder Strategy 2002-05*.

City of London Police (2002c), *City of London Police Annual Policing Plan*.

City of London Police (2002d), *Corporate Plan 2002-7*.

City of London Police (2003), *Crime – Terrorism* [online]. Available from: http://www.cityoflondon.police.uk/crime/terrorism.htm [accessed 3 January 2003].

City Security - Magazine of the City of London Crime Prevention associations in partnership with the City of London Police, 4/2000.

City Security - Magazine of the City of London Crime Prevention associations in partnership with the City of London Police, 9/2001.

Clark, D. (1996), *Urban World/Global City*, Routledge, London.

Clarke, L. and Ruddock, A. (1995), 'Is it a phoney peace?', *The Sunday Times*, 27 August, p. 12.

Clarke, R.V.G. (1992), *Situational Crime Prevention: Successful Case Studies*, Harrow and Heston, New York.

Clarke, R.V.G. and Mayhew, P. (eds.) (1980), *Designing Out Crime*, Home Office Reassert and Planning Unit, London, HMSO.

Clay, G. (1973), *Close-up: how to read the American City*, Pall Mall, London.

Cloke, P., Philo, C. and Sadler, D. (1991), *Approaching human geography*, Chapman, London.

Clutterbuck, R. (1990), 'Terrorism in Britain', *Wroxton Papers Series B*: The Changing Agenda of British Politics - B8.

Coaffee, J. (1996a), 'Terrorism, Insurance Rhetoric and the City of London', *Association of American Geographers 92 Annual Meeting*, Charlotte, North Carolina, 9-13 April 1996.

Coaffee, J. (1996b), 'Creating Images of Risk and security in Belfast and London', Annual Meeting of the Society for Risk Analysis-Europe - *Risk in a Modern Society: Lessons form Europe*, University of Surrey, Guildford, 3-5 June 1996.

Coaffee, J. (1996c), 'Beating the Bombers - Landscapes of Defence in London and Belfast', *28th International Geographical Congress*, The Hague, Netherlands, 4-10 August 1996.

Coaffee, J. (1997), 'The City of London's Ring of Steel - Panacea or placebo', *Fourth international seminar on urban form*, University of Birmingham, July 1997.

Coaffee, J. (2000a), 'Fortification, fragmentation and the threat of terrorism in the City of London in the 1990s', in Gold, J.R. and Revill G., *Landscapes of Defence*, Prentice Hall, London, pp. 114-129.

Coaffee, J. (2000b), *Risk, insurance and the making of the contemporary urban landscape*, unpublished PhD thesis, Oxford Brookes University.

Coaffee (2002), 'Recasting the Ring of Steel: Designing out terrorism in the City of London', paper presented at the *Cities as Strategic Sites Conference*, University of Salford, November.

Cockburn, C. (1977), *The local state: management of cities and people*, Pluto Press, London.

Coleman, A. (1984), 'Trouble in utopia: I. Design influences in blocks of flats', *Geographical Journal*, 150, pp. 351-62.

Coleman, A. (1985), *Utopia on trial: vision and reality in planned housing*, Hilary Shipman, London.

Collison, P. (1963), *The Cutteslowe Walls*, Faber and Faber, London.

Commissioner of the City of London Police (1994), 'The Future', *Citywatch - the City of London Police Magazine*, Summer, p. 1.

Compton, P.A., Murray, R.C. and Osborne, R. (1980), 'Conflict and its impact on the urban environment of Northern Ireland', in Enyedi, G. and Meszaros, J. (eds.) *Development of settlement systems*, Akademiai Kiado, Budapest, pp. 83-98.

Connon, H. and Durman, P. (1992), 'Bomb decision threatens property deal', *The Independent*, 7 December, p. 20.

Conzens. P., Hillier, D. and Prescott, G. (1999), 'Crime and the design of new build housing', *Town and Country Planning*, Vol. 68(7) pp. 231-3.

Conzens. P., Hillier, D. and Prescott, G. (2000), 'A tale of two cities', *Town and Country Planning*, Vol. 69(3), pp. 92-94.

Coogan, T.P. (1995), *The I.R.A.*, HarperCollins, London.

Cooke, P. (1988), 'Modernity, postmodernity and the city', *Theory Culture and Society*, Vol. 5(2-3), June.

Cooke, P. (ed.) (1995), *The rise of the rustbelt*, UCL Press, London.

Coone, T. (1993), 'Bishopsgate looks to Belfast', *Financial Times*, 30 April, p. 8.

Cooper, C. (1974), 'The house as a symbol of the self', in Lang, J. Burnette, C., Moleski, W. and Vachon, O. (eds.) *Designing for human behaviour*, Dowden, Hutchinson and Ross, Stroudsburg, pp. 130-46.
Corporation of London (1986), *The City of London Local Plan*, Department of Architecture and Planning.
Corporation of London (1993a), *The Way Ahead - Traffic and the Environment*, Draft Consultation Paper.
Corporation of London (1993b), *Security Initiatives*, Draft Consultation Paper.
Corporation of London (1994), *City Research Project: Final Report*, Corporation of London, London.
Corporation of London (1995), *City Research Project: Final Report* - the competitive position of London's financial services, Corporation of London, London.
Corporation of London (2001a), Press release, 12 September.
Corporation of London (2001b), Press release, 14 November.
Cottle, S. (1998), 'Ulrich Beck, risk society and the media', *European Journal of Communication*, Vol. 13(1), pp. 5-32.
Council of Europe (1993/1998), *The European Urban Charter*, Council of Europe Press.
Counsell, G., Thapar, N. and Kelsey, T. (1993), 'City's role secure after IRA bombing', *The Independent*, 26 April, p. 22.
Coward, M. (2002), 'Community as heterogeneous ensemble: Mostar and multiculturalism', *Alternatives,* 27, pp. 29-66.
Crabbe, H. (1996), 'Can traffic control clean the air?', *Air Health Strategy,* August, pp. 8-9.
Craven, E. (1975), 'Private residential expansion in Kent', in Pahl, R.E. *Whose City?*, Penguin, Harmondsworth, pp. 124-44.
Crawford, M. (1995), 'Contesting the public realm: struggles over public space in Los Angeles', *Journal of Architectural Education*, 49, pp. 4-9.
Cresswell, T. (1992), 'The crucial "where" of graffiti: a geographical analysis to reactions to graffiti in New York', *Environment and Planning D Society and Space*, 10, pp. 329-44.
Crilley, D. (1993a), 'Architecture as advertising: constructing the image of redevelopment', in Kearns, G. and Philo, C. (eds.) *Selling Places: The city as cultural capital - past and present*, Pergamon Press, Oxford, pp. 231-52.
Crilley, D. (1993b), 'Megastructures and urban change: aesthetics, ideology and design', Knox, P. (ed.) (1993), *The Restless Urban Landscape*, PrenticeHall, New Jersey, pp. 127-64.
Cullen, J. and Knox, P. (1981), 'The triumph of the eunuch: planners, urban managers and the suppression of political opposition', *Urban Affairs Quarterly*, Vol. 16, pp. 149-172.
Curphey, M. (2003), 'The chancellor's emergency measures', *The Guardian*, 26 February.
Cuthbert, A. (1995), 'Under the volcano: postmodern space in Hong Kong', in Watson, S. and Gibson, K. (eds.) *Postmodern cities and spaces*, Blackwell, Oxford, pp. 138-48.
Dandeker, C. (1990), *Surveillance, power and modernity*, Polity Press, Cambridge.
Daniels, P.W. and Bobe, J.M. (1992), 'Office building in the City of London: a decade of change', *Area*, Vol. 24(3), pp. 253-8.
Daniels, S. (1993), *Fields of vision: landscape imagery and national identity in England and the United States*, Princeton University Press, Princeton.
Davidson-Smith, G. (1990), *Combating Terrorism*, Routledge, London.
Davies, S. (1996a), *Big Brother – Britain's web of surveillance and the new technological* order, Pan Books, London.
Davies, S. (1996b), 'The Case Against: CCTV should not be introduced', *International Journal of Risk, Security and Crime Prevention*, Vol. 1, No.4, pp. 327-31.
Davis, M. (1987), 'Chinatown, part two? The internationalisation of Los Angeles', *New Left Review*, 164, p. 65.

Davis, M. (1990), *City of Quartz - excavating the future of Los Angeles*, Verso, London.
Davis, M. (1992), *Beyond Blade Runner: urban control - the ecology of fear*, Open Magazine Pamphlet series, Westfield, New Jersey.
Davis, M. (1995), 'Fortress Los Angeles: the militarization of urban space', in Kasinitz, P. (ed.) *Metropolis - centre and symbol of our times*, Macmillian, London.
Davis, M. (1998), *Ecology of Fear: Los Angeles and the imagination of disaster*, Metropolitan Books, New York.
Davis, M. (2001), 'The Future of Fear', *City*, Vol. 5(3), pp. 388-90.
Dawson, G.M. (1984), 'Defensive planning in Belfast', *Irish Geography*, 17, pp. 27-41.
De Souza, M.J.L. (1996), 'The fragmented metropolis - drug traffic and its territoriality in Rio De Janeiro', *Geographische Zeitschrift*, Vol. 83(3-4), pp. 238-49.
Dear, M. (1995), 'Prolegomena to a postmodern urbanism', in Healy, P., Cameron, S., Davoudi, S., Graham, S. and Madanipour, A. (eds.) *Managing Cites - the new urban context*, Wiley, Chichester.
Dear, M. and Flusty, S. (1998), 'Postmodern urbanism', *Annals of the Association of American Geographers*, Vol. 88(1), pp. 50-72.
Dear, M.J. (1999), *The postmodern urban condition*, Blackwell, Oxford.
Demko, J., and Wood, W.B. (eds.) (1994), *Reordering the world: Geopolitical perspectives on the 21st century*, West View Press, Boulder.
Diacon, S.R. and Carter, R.L. (1995), *Success in insurance*, John Murray, London.
Dickson, G.C.A. and Steele, J.T. (1995), *Introduction to insurance*, Pitman Publishing, London.
Dillon, D. (1994), 'Fortress America', *Planning*, 60, pp. 8-12.
Dillon, M. (1994), *25 years of terror - the IRA's war against the British*, Bantam Books, London, pp. 355-68.
Dingemans, D. (1979), 'Redlining and mortgage lending in Sacramento', *Annals of the Association of American Geographers*, Vol. 69(2), pp. 225-37.
Ditton, J. (1996), *Does closed circuit television prevent crime?*, Scottish Centre for Criminology, Edinburgh.
Do or Die (1999), Number 8 – *The special pre millennium tension issues*.
Dobbs, P. (1990), 'The business of terror', *The Journal - the Chartered Insurance Institute*, January, pp. 16-18.
Dobkin, B.A. (1992), *Tales of terror: television news and the construction of the terrorist threat*, Praeger, New York.
Doeksen, H. (1997), 'Reducing crime and the fear of crime by reclaiming New Zealand's suburban street', *Landscape and Urban Planning*, 39, pp. 243-52.
Domosh, M. (1987), 'Imagining New York's first skyscrapers 1875-1910', *Journal of Historical Geography*, 13, pp. 233-48.
Donegan, L. (1993), 'City traffic ban to help fight bombers', *The Guardian*, 22 May, p. 9.
Doornkamp, J.C. (1995), 'Perception and reality in the provision of insurance against natural perils in the UK', *Transactions of the Institute of British Geographers*, Vol. 20(1), pp. 68-80.
Douglas, M. (1994), *Risk and Blame - essays in cultural theory*, Routledge, London.
Douglass, W.A. and Zulaika, J. (1998), 'On terrorism discourse: reply to Greenwood', *Current Anthropology*, 39(2), pp. 265-6.
DTI Press Notice - Michael Heseltine announcement on insurance cover for terrorism, 21 December 1992.
Duffy, F. and Henney, A. (1989), *The Changing* City, Bulstrode Press, London.
Duncan, J. and Ley, D. (eds.) (1993), *Place/culture/representation*, Routledge, London.
Durham, P. (1995), 'Villains in the frame', *Police Review*, 20 January, pp. 20-21.

Durhan, P. (1993), 'Cost will exhaust pool of insurance money', *The Independent*, 26 April, p. 2.

Dymski, G.A. (1995), 'The theory of bank redlining and discrimination - an exploration', *Review of the Black Political Economy*, Vol. 23(3), pp. 37-74.

Eade, J. (1997), *Living the Global City*, Routledge, London.

Edney, J.J. (1976), 'The psychological role of property rights in human behaviour', *Environment and Planning*, 8A, pp. 811-22.

Ekbolm, P. (1995), 'Less crime by design', *Annals AAPSS* (No.539), May, pp. 114-29.

El-Sabh, M.I. and Murty, T.S. (eds.) (1988), *Natural and Man-made Hazards*, D.Reidel, Dordrecht.

Ellin, N. (1996), *Postmodern urbanism*, Blackwell, Oxford.

Ellin, N. (ed.) (1997), *Architecture of fear*, Princeton Architectural Press, New York.

Elliot, B. and McCrone, D. (1975), 'Property relations in the city: the fortunes of landlordism', in Harloe, M. (ed.) *Proceedings of the conference on urban change and conflict*, pp. 31-36.

Elliot, C. and Mackay, A. (1993), 'Bankers defy government to seek tougher security', *The Times*, 5 June, p. 2.

Engels, B. (1994), 'Capital flows, redlining and gentrification - the pattern of mortgage lending and social-change in Glebe, Sydney, 1960-1984', *International Journal of Urban and Regional Research*, Vol. 18(4), pp. 628-57.

Ericson, R.V. and Haggerty, K.D. (1997), *Policing the risk society*, University of Toronto Press, Toronto.

Eriksen, E.G. (1980), *The territorial experience: human ecology as symbolic interaction*, University of Texas Press, Austin.

Esser, A.H. (ed.) (1971), *Behaviour and environment*, Plenum, New York.

Evans, D.J. (1995), *Crime and policing: spatial approaches*, Avebury, Hants.

Ewald, E. (1986), *L'Etat providence*, Paris.

Ewald, F. (1993), 'Two infinities of risk', in Massumi, B. (ed.) *The politics of everyday fear*, University of Minnesota Press, Minneapolis, pp. 221-28.

Fainstein, S.S. (1994), *The city builders: property, politics and planning in London and New York*, Blackwell, Oxford.

Fainstein, S.S. (2000), 'New directions in planning theory', *Urban Affairs Review*, Vol. 35(4), pp. 451-78.

Farish, M. (2002), 'Another Anxious Urbanism: Simulating Defence and Disaster in Cold War America', paper presented at the *Cities as Strategic Sites conference*, University of Salford, November.

Faulsh, P. (1994), *'Inner city blues'*, *Post Magazine*, 17 November.

Featherstone, M. (1988), 'In Pursuit of the postmodern: an introduction', *Theory, Culture and Society*, Vol. 5(2-3), June.

Feldstein, M. (1994), 'Insurance redlining- hitting the poor where they live', *Nation*, Vol. 258(13), p. 450.

Finch, P. (1996), 'The fortress city is not an option', *The Architects' Journal*, 15 February, p. 25.

Fisher, B.S. (1993), 'Community response to crime and fear of crime: the socio economic of crime', in Forst, B. (ed.) *Crime and Justice*, Sharp, New York, pp. 177-207.

Fisk, R. (1997), 'Back to Belfast', *The Independent*, 12 February, p. 12.

Flaschsbart, P.G. (1969), 'Urban territorial behaviour', *Journal of the American Institute of Planning*, 35, pp. 412-6.

Flusty, S. (1994), *Building paranoia - the proliferation of interdictory space and the erosion of spatial justice*, Los Angeles Forum for Architecture and Urban Design, 11.

Forbes, R.J. (1965), *Irrigation and power*, Bull Leiden.

Ford, J. (1975), 'The role of the building society manager in the urban stratification system', *Urban Studies* 12, pp. 295-302.

Ford, R. (1993a), 'Security will be tightened for City prestige targets', *The Times*, 26 April, p. 2.

Ford, R. (1993b), 'Police resist demands for City ring of steel', *The Times*, 27 April, p. 3.

Freedman, J.L. (1975), *Crowding and behaviour*, Freeman, San Francisco.

Freidmann, J.P. (1986), 'The world city hypothesis', *Development and Change*, 17, pp. 69-74.

Friedman, D. (1998), 'The ecology of Mike Davis', *Los Angeles Downtown News*, 9 October.

Fumagalli, V. (1994), *Landscapes of fear: perceptions of nature and the city in the Middle ages* (translated by S. Mitchell), Polity Press, Cambridge.

Furedi, F. (2002), *Culture of Fear: Risk taking and the morality of low expectation*, Continuum International Publishing, New York.

Fyfe, N.R. (1997), 'Commentary on policing space', *Urban Geography*, Vol. 18(5), pp. 389-91.

Fyfe, N.R. (1998), '*Images on the street*', Routledge, London.

Fyfe, N.R. and Bannister, J. (1996), 'City watching: close circuit television surveillance in public spaces', *Area*, 28.1, March, pp. 37-46.

Fyfe, N.R. and Bannister, J. (1998), 'The eyes upon the street: close circuit television surveillance and the city', in Fyfe, N. R. (1998), '*Images on the Street*', Routledge, London.

Gabor, T. (1981), 'The crime displacement hypothesis; and empirical examination', *Crime and Delinquency*, Vol. 26, pp. 390-404.

Gabor, T. (1990), 'Crime displacement and situational prevention: towards the development of some principles', *Canadian Journal of Criminology*, Vol. 32, pp. 41-74.

Garvey, G. (1993), 'City security boosted in war on terrorism', *Evening Standard*, 29 April, p. 5.

Geake, E. (1993), 'The electronic arm of the law', *New Scientist*, 8 May, pp. 19-20.

Gearty, C. (1991), *Terror*, Faber and Faber, London.

Giddens, A. (1984), *The constitution of society*, University of California Press, Berkeley.

Giddens, A. (1981), *A contemporary critique of historical materialism, Vol. 1*, Macmillan, London.

Giddens, A. (1985), *The nation state and violence*, Polity Press Cambridge.

Giddens, A. (1990), *The consequences of modernity*, Polity press, Cambridge.

Giddens, A. (1991), *Modernity and self identity: self and society in the late modern age*, Polity Press, Cambridge.

Giddens, A. (1992), 'Commentary on the reviews', *Theory, Culture and Society*, Vol. 9, pp. 171-74.

Giddens, A. (1999), *Runaway world: how globalisation is reshaping our lives*, Profile Books, London.

Giles, M. (1994), 'On shaky ground - insurers are being asked to deal with unprecedented risks, creating pressure on governments to foot more of the bill. That would be a mistake', *The Economist*, 3 December.

Glaeser, E.L. and Shapiro, J.M. (2002), 'Cities and Warfare: The impact of Terrorism on Urban Form', *Journal of Urban Economics*, 51, pp. 205-24.

Glaister, S. and Travers, T. (1993), *Meeting the transport needs of the City*, Corporation of London.

Glassner, B. (2002), *The Culture of Fear: Why Americans are afraid of the wrong things*, Basic Books, New York.

Gloyn, W.J. (1993), *Insurance against terrorism*, Witherby, London.

Gold, J.R. (1980), *An introduction to behavioural geography*, Oxford University Press, Oxford.

Gold, J.R. (1982), 'Territoriality and human spatial behaviour', *Progress in Human Geography*, 6, pp. 44-67.

Gold, J.R. (1994), 'Locating the message: place promotional image communication', in Gold, J.R. and Ward, S.V. (eds.) *Place Promotion*, Wiley, Chichester, pp. 19-37.

Gold, J.R. (1997), *The Experience of modernism: modern architects and the future city, 1928-1953*, E. & F.W. Spon, London.

Gold J.R and Coaffee, J. (1998), 'Tales of the city: understanding urban complexity through the medium of concept maps', *Journal of Geography in Higher Education*, Vol. 22(3), pp. 285-96.

Gold, J.R. and Revill G. (1999), 'Landscapes of defence', *Landscape Research*, Vol. 24(3), pp. 229-39.

Gold, J.R. and Ward, S.V. (eds.) (1994), *Place Promotion*, John Wiley, Chichester.

Gold, R. (1970), 'Urban violence and contemporary defensive cities', *Journal of the American Institute of Planning*, 36, pp. 146-59.

Goodey, B. and Gold, J.R. (1987), 'Environmental Perception: the relationship with urban design', *Progress in Human Geography*, pp. 126-32.

Goodwin, M. (1993), 'The city as commodity: the contested spaces of urban development', in Kearns, G. and Philo, C. (eds.) *Selling places: The city as cultural capital - past and present*, Pergamon Press, Oxford, pp. 145-162.

Goodwin, M. (1995), 'Governing the spaces of difference: regulation and globalisation in London', a paper presented at the *10th Urban Change and Conflict Conference*, Royal Holloway, University of London, 5-7 September.

Gorman, E. and Hicks, J. (1995), 'IRA killings could herald a return to full scale war', *The Times*, 29 December, p. 2.

Goss, J. (1997), 'Representing and re-presenting the contemporary city', *Urban Geography*, Vol. 18(2), pp. 180-88.

Gottdiener, M. (1994), *The social production of space*, University of Texas Press, Austin.

Gottdiener, M. (1995), *Postmodern semiotics: material culture and the forms of modern life*, Blackwell, Oxford.

Gottdiener, M. (2000), *The Theming of America: American Dreams, Media Fantasies, and Themed Environments*, Westview Press, Boulder.

Graham, S. (1995), *Towns on the television: CCTV surveillance in British towns and cities*, Department of Town and Country Planning, University of Newcastle, Working Paper 50).

Graham, S. (1999), 'The eyes have it – CCTV as the fifth utility', *Town and Country Planning*, Vol. 68(10), pp.312-5.

Graham, S. (2001), 'In a moment: on global mobilities and the terrorised city', *City*, Vol. 5(3), pp. 411-15.

Graham, S. (2002a), 'Special collection: Reflections on Cities. September 11th and the 'war on terrorism' – one year on', *International Journal of Urban and Regional Research*, Vol. 26(3), pp. 589-90.

Graham, S. (2002b), 'CCTV: the stealthy emergence of a fifth utility?', *Planning Theory and Practice*, Vol. 3(2), pp. 237-41.

Graham, S. (2002c), Urbanizing War/Militarizing Cities: The city as strategic site', *Archis*, 3, pp. 25-33.

Graham, S. (2003), 'Lessons in Urbicide, *New Left Review*, 19, pp. 63-77.

Graham, S. and Marvin, S. (1996), *Telecommunications and the City - electronic spaces, urban places*, Routledge, London.

Graham, S. and Marvin, S. (2001), *Splintering Urbanism – networked infrastructures, technological mobilities and the urban condition*, Routledge, London.

Granovettor, M. (1985), 'Economic action, social structure and embededness', *American Journal of Sociology*, 91, pp. 481-510.

Gray, F. (1976), 'Selection and allocation in council housing', *Transactions of the Institute of British Geographers*, Vol. 1(1), pp. 34-46.

Greater London Assembly (2000), *Scrutiny of the Mayor's Proposals for Congestion Charging – Report of Findings*.

Greenbie, G.G. (1982), 'The landscape of social symbols', *Landscape*, Vol. 7(3), pp. 2-6.

Greer, D.S, and Mitchell, V.A, (1982), *Compensation for criminal damage to property*, SLS Legal publications, Belfast.

Griffith, I.L. (1999), 'Organized crime in the Western hemisphere: content, context, consequence and countermeasures', *Low Intensity Conflict and Law Enforcement*, 8, pp. 1-33.

Griffiths, C. (1995), *Tackling fear of crime: a starter kit*, HMSO, London.

Grimond, M. (1998), 'Schroder doubles up against disaster fears', *London Evening Standard Online*, 26 March.

Grogger, J. and Weatherford, M.S. (1995), 'Crime, policing and the perception of neighbourhood safety', *Political Geography*, Vol. 14(6-7), pp. 521-41.

Gunn, C. (1992), 'Heseltine backs down on terror bomb cover', *Today*, 21 December, p. 24.

Guskind, R. (1995), 'Thin red line', *National Journal*, Vol. 21(43), pp. 2639-43.

Gusmaroli, D. (1993), 'Steel ring must stay', *City Post*, 15 July, p. 1.

Guy, J. (1996), 'IRA attack shatters terrorism policy plans', *Post Magazine*, 20 June.

Guy, J. (1997), 'Insurers still feeling the force of the blast', *Post Magazine*, 13 February.

Guy, S. and Harris, R. (1997), 'Property in a global risk society; towards marketing research', *Urban Studies* Vol. 34(1), pp. 125-40.

Habermas, J. (1984), *The theory of communicative action: Vol. 1 reason and rationalisation of society*, Polity Press, Cambridge.

Hakim, S. and Rengert, G.F. (1981), *Crime spillover*, Sage, Beverly Hills.

Hall, E.T. (1966), *The hidden dimension*, Doubleday, New York.

Hall, J.R. (1994), 'Book review', *Sociological Review*, Vol. 42(2), May, pp. 344-6.

Hall, P. (1966/1977), *The World Cities*, Weidenfield and Nicholson, London.

Hall, P. (1996), *Cities of tomorrow*, Blackwell, Oxford.

Hall, P. (2001) 'The unthinkable event that may doom high-rise', *Regeneration and Renewal*, 21 September, p.14.

Hall, T. (1998), *Urban Geography*, Routledge, London.

Hall, T. and Hubbard, P. (1996), 'The entrepreneurial city: new urban politics, new urban geography?', *Progress in Human Geography*, Vol. 20(2), pp. 153-174.

Hamilton, A. (1986), *The Financial Revolution*, Penguin, Middlesex.

Hammnett, C. (1996), 'Classics in human geography revisited - commentary 1', *Progress in Human Geography*, Vol. 20(3), pp. 373-5.

Harrigan, J. and Martin P. (2002), 'Terrorism and the resilience of cities', *FRBNY Economic Policy Review*, November, pp. 97-116.

Harris, J, and Thane, P. (1984), 'British and European bankers, 1880-1914: an aristocratic bourgeoisie?', in Thane, P., Crossick, G. and Floud, R. *The Power of the Past: Essays for Eric Hobsbawn*, Cambridge University Press, Cambridge, pp. 215-34.

Harrison, B. (1994), *Lean and mean: the changing landscape of corporate power in the age of flexibility*, Basic Books, New York.

Harvey, D. (1974), 'Class monopoly rent, finance capital and urban revolution', *Regional Studies*, 8, pp. 239-55.

Harvey, D. (1985), *The urbanisation of capital*, Blackwell, Oxford.

Harvey, D. (1989), 'From managerialism to entrepreneuarialism and the transformation of urban governance in late capitalism', *Geografiska Annaler*, 71, pp. 3-17.

Harvey, D. (1990), *The Condition of Postmodernity*, Blackwell, Oxford.

Harvey, D. (1996), *Justice nature and the geography of difference*, Blackwell, Oxford.

Harvey, D. (1997), 'The new urbanism and the communitarian trap', *Harvard Design Magazine,* Winter/Spring.

Hassan, I. (1985), 'The culture of postmodernism', *Theory, Culture and Society*, Vol. 2(3), pp. 119-132.

Haynes, S.E. (1995), 'Beating the clock on terrorism', *Public Risk*, October, pp. 8-9.

Healey, P. (1992), 'An institutional model of the development process', *Journal of Property Research*, 9, pp. 33-44.

Healey, P. (1996), 'The communicative turn in planning theory and its implication for spatial strategy formation', *Environment and Planning B: Planning and Design*, 23, pp. 217-34.

Healey, P. (1997), *Collaborative planning - shaping places in fragmented societies*, Macmillan, Basingstoke.

Healey, P. (1998), 'Building institutional capacity through collaborative approaches to planning', *Environment and Planning A*, 30, pp. 1531-1546.

Healey, P. (1999), 'Institutional analysis, communicative planning and shaping places', *Journal of Planning Education and Research*, 19, pp. 111-21.

Healey, P., and Barrett, S. (1990), 'Structure and agency in land and property development processes: some ideas for research', *Urban Studies*, Vol. 27(1), pp. 89-104.

Healey, P., Cameron, S., Davoudi, S., Graham, S. and Madanipour, A. (eds.) (1995), *Managing Cities*, John Wiley, London.

Healey, P., Madanipour, A. and de Magalhaes, C. (1999), 'Institutional capacity, social capital and urban governance', paper to a seminar on *Institutional capacity, social capital and urban governance*, Newcastle, April.

Hebbert, M. (1998), *London: more by fortune than design*, Wiley, Chichester.

Henneberry, J. (1983), 'Who provides the money', *Planner*, Vol. 69(6), pp. 200-01.

Herbert, D. and Thomas, C. (1982), *Urban Geography*, John Wiley, Chichester.

Herbert, S. (1996), 'The normative ordering of police territoriality - making and marking space with the Los Angeles police department', *Annals of the Association of American Geographers*, Vol. 86(3), pp. 567-582.

Herbert, S. (1997a), *Policing space: territoriality and the Los Angeles police department*, University of Minnesota Press, Minneapolis.

Herbert, S. (1997b), 'On prolonging the conversation: some correctives and continuities', *Urban Geography*, Vol. 18(5), pp. 398-402.

Herbert, S. (1998), 'Policing contested space: on patrol in Smiley and Hauser', in Fyfe, N. R. (ed.) (1998), *Images on the* Street, Routledge, London.

Hermon, J. (1990), 'The police, the media and the reporting of terrorism', in Alexander, Y., and Latter, R. (eds.) *Terrorism and the media - dilemmas for government, journalists and the public*, Brassey's, London, pp. 37-42.

Hill, M. (1995), 'Beating the bombers', *Police Review*, 17 February, pp. 14-16.

Hillier, T. (1994), 'Bomb attacks in city centres' [online]. Available from: *http://www.emergency.com/carbomb.htm.*

Hirst, P. and Thompson, G. (1996), *Globalisation in question: the international economy and the possibilities of governance*, Polity Press, Cambridge.

Hobsbawn, E. (1983), 'Introduction: inventing tradition', in Hobsbawn, E. and Ranger, T. *The Invention of Tradition*, Cambridge University Press, Cambridge.

Hodgson, G.M. (1988), *Economics and Institutions*, Polity Press, Cambridge.

Hoffman, B and Hoffman, D.K. (1996), 'Chronology of international terrorism 1995', *Terrorism and Political Violence*, Vol. 8(3), pp. 87-127.

Hoffman, B. (1998), *Inside Terrorism*, Indigo, London.

Holcomb, B. (1993), 'Revisoning place: De- and re-constructing the image of the industrial city', in Kearns and Philo (eds.) (1993), *Selling places: the city as cultural capital, past and present*, Pergamon Press, Oxford, pp. 133-143.

Home Office (1994), *CCTV: looking out for you*, HMSO, London.

Honess, T. and Charman, E. (1992), '*Closed circuit television in public places*', Crime Prevention Unit Paper 35, Home Office, London.

Hope, T. and Sparkes, R. (eds.) (2000), *Crime, risk and insecurity*, Routledge, London.

Horne, C.J. (1996), 'The case for: should CCTV be introduced', *International Journal of Risk, Security and Crime Prevention,* Vol. 1(4), pp. 317-326.

Howard, H.E. (1920), *Territory and bird life*, Murray, London.

Hoyt, H. (1939), *The structure and growth of residential neighbourhoods in American cities*, US Federal Housing Administration, Washington.

Hubbard, P. (1996), 'Transformation of Birmingham's urban landscape', *Geography*, Vol. 81(1), pp. 26-36.

Huber, P. and Mills M. (2002), 'How technology will defeat terrorism', *City Journal*, Vol. 12(1), pp. 24-34.

Hudson, J.D. (1973), *The effects of the civil disturbances in Northern Ireland upon shopping centre patterns and other patterns of movement with particular refenrce to the city centre*, unpublished undergraduate dissertation, Geography Department, Queens University Belfast.

Hudson, R. (1995), 'Institutional change, cultural transformation, and economic regeneration: myths and realities from Europe's old industrial areas', in Amin, A. and Thift, N. (eds.) *Globalization, Institutions and Regional Development in* Europe, Oxford University Press, Oxford, pp. 196-216.

Hunter, A. (1995), 'Local knowledge and local power', in Hertz, R. and Imber, J.B. (eds.) (1995), *Studying elites using qualitative methods*, Sage, Thousand Oaks.

Hyett, P. (1996), 'Damage limitation in the age of terrorism', *The Architects' Journal*, 15 February, p. 29.

Imrie, R. (1996), 'Review of Healey et al (eds.), Managing Cities: the new Urban Context', *European Urban and Regional Studies*, 3, pp. 374-5.

Imrie, R. and Raco, M. (1999), 'How new is the new local governance? Lessons from the United Kingdom', *Transactions of the Institute of British Geographers*, Vol. 24(1), pp. 45-63.

Imrie, R. and Thomas, H. (1995), 'Changes on local governance and their implications for urban policy evaluation', in Hambleton, R. and Thomas, H (eds.), *Urban policy evaluation: challenge and change*, Paul Chapman, London, pp. 123-138.

Investors Chronicle (1996), 'The Man who helped to build a business district', 5 July.

Jack, A. (1993), 'City critical of government after bomb', *Financial Times*, 1 May, p. 6.

Jackson, P. (1992), *Maps of meaning*, Routledge, London.

Jackson, P. (1999), 'Postmodern urbanism and the ethnographic void', *Urban Geography*, Vol. 20 (5), pp. 400-402.

Jacobs, J. (1961, 1984 edition), *The Death and Life of Great American Cities*, Peregrine, London.

Jacobs, J.M. (1992), 'Cultures of the past and urban transformations: the Spitalfields market development in east London', in Anderson, K and Gale F. (eds.), *Inventing places: studies in cultural geography*, Longman, Cheshire, pp. 194-314.

Jacobs, J.M (1993), 'The Battle at Bank Junction: the contested iconography of capital', in Corbridge, S., Martin, R. and Thrift, N. (eds.), *Money, Power and Space*, Blackwell, Oxford, pp. 356-382.

Jacobs, J.M. (1994), 'Negotiating the heart', *Environmental and Planning D: Society and Space*, 12, pp. 751-72.

Jacoby, C. (1993), 'Building a ring of steel', *Estates Gazette*, 27 November, pp. 99-102.

Jarman, N. (1993), 'Intersecting Belfast', in Bender, B. (ed.) *Landscape - politics and perspectives*, Berg, Oxford, pp. 107-138.

Jeffery, C.R. (1971), *Crime prevention through environmental design*, Sage, Beverly Hills.

Jencks, C. (1977), *The language of post-modern architecture*, Academy Editions, London.

Jessop, B. (1997), 'The entrepreneurial City: Re-imaging localities, redesigning economic governance, or restructuring capital?', in Jewson, N. and MacGregor, S., *Transforming cities - contested governance and new spatial divisions*, Routledge, London, pp. 28-41.

Jewson, N. and MacGregor, S. (1997), *Transforming cities - contested governance and new spatial divisions*, Routledge, London.

Johnson, L.C. (1997), 'The fall of terrorism', *Security Management (Internet edition)*, April.

Johnson, R.J. (1997), *Geography and geographers (fifth edition)*, Arnold, London.

Johnstone, C. (2002), 'Realising the fifth utility?', *Town and Country Planning*, Vol. 71(11), pp. 286-9.

Jones, C., and Maclennan, D. (1987), 'Building societies and credit rationing - an empirical examination of redlining', *Urban Studies*, Vol. 24, pp. 205-16.

Jones, G. (1996), 'Clinton call for alliance to combat terrorism', *Electronic Telegraph*, Issue 416, 28 June.

Jones, M.C. and Lowrey, K.J (1995), 'Street barriers in American cities', *Urban Geography*, Vol. 16(2), pp. 112-22.

Jones, T. (1993), 'Company chiefs want steel gates over roads', *The Times*, 27 April, p. 3.

Jones, T. and Newburn, T. (1998), *Private security and public policing*, Clarendon Press, Oxford.

Jordan, T.G., Domosh, M. and Rowntree, L. (1997), *The human mosaic - a thematic introduction to cultural geography*, Longman, London.

Juergensmeyer, M. (1997), 'Terror mandated by God', *Terrorism and Political Violence*, Vol. 9(2), pp. 16-23.

Kantor, A.C. and Nystuen, J.D. (1982), 'De Facto redlining: a geographic view', *Economic Geography*, Vol. 58(4), pp. 309-28.

Kaplan, S. and Kaplan, R. (eds.) (1978), *Humanscape: The environments for people*, Duxbury, Wadsworth (Mass).

Kates, R.W. (1962), *Hazard and choice perception in flood plain management*, Paper 78, Department of Geography, University of Chicago.

Katz, C. (1993), 'Reflections while reading City of Quartz by Mike Davis', *Antipode*, 25(2), pp. 159-63.

Kay, J. and Lewthwaite, J. (1993), 'Untitled', *The Sun*, 26 April, p. 4.

Kearns and Philo (1993), *Selling places: the city as cultural capital, past and present*, Pergamon Press, Oxford.

Keating, M. (1993), 'The politics of local economic development', *Urban Affairs Quarterly*, 28, pp. 373-96.

Keegan, J. (1993), *A history of warfare*, Hutchinson, London.

Kellner, D. (2002), 'September 11, Social Theory and Democratic Politics', *Theory, Culture and Society*, Vol. 19(4), pp. 147-59.

Kelly, O. (1994a), 'By all means necessary', *Police Review*, 15 April, pp. 14-16.

Kelly, O. (1994b), 'The IRA threat to the City of London', *Policing*, p. 10.

Kenny, J. (1992), 'Portland's comprehensive plan as text: the Fred Meyer case and the politics of reading', in Barnes T.J. and Duncan J.S., *Writing Worlds: discourse, text and metaphor in the representation of the landscape*, Routledge, London, pp. 176-92.

King, A. (1990), 'Architecture, capital and the globalization of culture', *Theory, Culture and Society*, Vol. 7, pp. 397-411.

King, A.D. (1990), *Global cities: post imperialism and the internationalisation of London*, Routledge, London.

Kirk, W. (1963), 'Problems of geography', *Geography*, 48, pp. 357-71.

Kirsch, S (2001), 'The world risk society and the policing of terror: notes on insurance, internationalism and democracy', paper prepared for UNC-Chapel Hill progressive Faculty Network teach in, *'Understanding Terror: What is War? What is Peace?*, 1 October 2001.

Kleinman, M. (1999), 'The business sector and the governance of London', Paper presented at the *European Cities in Transformation Conference*, Université de Marne-la-Vallée, Paris, 22-23 October.

Knox, P. (1989), *Urban social geography - an introduction*, Longman Group, London.

Knox, P. (1984), 'Symbolism, styles and settings: the built environment and the imperatives of urbanised capital', *Architecture and Behaviour*, Vol. 2, pp. 107-122.

Knox, P. (1987), 'The social production of the built environment: architects, architecture and the post-modern city', *Progress in Human Geography*, Vol. 11, pp. 354-77.

Knox, P. (ed.) (1993), *The restless urban landscape*, PrenticeHall, New Jersey.

Knox, P.L. (1996), 'Globalization and the world city hypothesis', *Scottish Geographical Magazine*, 112, pp. 124-6.

Knox, P.L. (1993), 'The restless urban landscape: economic and sociocultural change and the transformation of Metropolitan Washington, DC', *Annals of the Association of American Geographers*, Vol. 83(2), pp. 181-209.

Kosstoff, J. and Thompson, T. (1994), 'Policing the peace', *Time-Out*, 7-14 September, pp. 12-13.

Kotler, P., Haider, D.H. and Rein, I.J. (1994), *Marketing places: attracting investment, industry and tourism to cities, states and nations*, Free Press, New York.

Kumar, K. (1995), *From post-industrial to post-modern society*, Blackwell, Oxford.

Kunstler, J. and Salingaros, N.A. (2001). *The End of Tall Buildings*, published electronically by PLANetizen [online], 17 September. Available from http://www.planetizen.com/.

Kynaston. D. (2001), *The City of London, Volume IV: a club no more*, Chatto and Windus, London.

Ladbury, A. (1995), 'U.K risk mangers seek peace dividend', *Business Insurance*, 9 January, pp. 19-21.

Lake, R.W. (1999), 'Postmodern urbanism?', *Urban Geography*, Vol. 20(5), pp. 393-5.

Lapper, R. (1993a), 'Insurance costs expected to rise', *Financial Times*, 26 April, p. 7.

Lapper, R. (1993b), 'City faces rise in terror premiums', *Financial Times*, 29 May, p. 6.

Laqueur, W. (1996), 'Post-modern terrorism', *Foreign Affairs*, Vol. 75(5), pp. 24-36.

Lash, S. (1990), *Sociology of postmodernism*, Routledge, London.

Lash, S., Szerszynski, B. and Wynne, B. (1996), *Risk, environment and a new modernity - towards a new ecology*, Sage, London.

Lash. S. and Urry, J. (1994), *Economies of signs and spaces*, Sage, London.

Lash, S. and Wynne, B. (1992), 'Introduction' in Beck, U. (1992), *Risk Society - towards a new modernity*, Sage, London, pp. 1-8.

Lauria, M. (1982), 'Selective urban regeneration: a political economic perspective', *Urban Geography*, Vol. 3(3), pp. 224-39.

Lauria, M. and Knopp, L. (1985), 'Towards an analysis of the role of gay communities in the urban renaissance', *Urban Geography*, 6, pp. 152-169.

Leader, S.H (1997), 'The rise of terrorism', *Security Management (Internet edition)*, April.

Lee, R.M. (1993), *Doing research on sensitive topics*, Sage, London.

Lee, R.M. and Renzetti, C.M. (1990), 'The problems of researching sensitive topics: an overview and introduction', *American Behavioural Scientist*, 33, pp. 510-33.

Lees, L. (1998), 'Urban resistance and the street: spaces of control and contestation', in Fyfe, N.R. (1998), *Images on the Street*, Routledge, London.

Leiss, W. (1994), 'Book Review', *Canadian Journal of Sociology*, Vol. 19(4).

Leonard, S. (1982), 'Managerialism: a concept of relevance?', *Progress in Human Geography,* 7, pp. 212-23.

Ley, D. (1996), 'Urban geography and cultural studies', *Urban Geography*, Vol. 17(6), pp. 475-7.

Ley, D. (1974), *The black inner city as frontier outpost: images and behaviour of a Philadelphia neighbourhood*, Monograph No.7, Association of American Geographers, Washington D.C.

Ley, D. (1983), *A social geography of the city*, Harper and Row, New York.

Ley, D. (1988), 'From urban structure to urban landscape', *Urban Geography,* Vol. 9(1), pp. 98-105.

Ley, D. (1989), 'Modernism, postmodernsim and the struggle for place', in Agnew and Duncan (eds.), *The Power of Place*, Unwin Hyman, London, pp. 44-65.

Ley, D. and Cybriwsky, R. (1974), 'Urban graffiti as territorial markers', *Annals of the Association of American Geographers*, 64, pp. 491-505.

Leyshon, A. and Thrift, N. (1994), 'Access to financial services and financial infrastructural withdrawal - problems and policies', *Area*, Vol. 26(3), pp. 268-75.

Leyshon, A. and Thrift, N. (1995), 'Geographies of financial exclusion: financial abandonment in Britain and the United States', *Transactions of the Institute of British Geographers*, Vol. 20(3), pp. 312-41.

Leyshon, A. and Thrift, N. (1997), 'Spatial financial flows and the growth of the modern city', *International Social Science Journal*, 151, 41-53.

Li, Y. (1997), 'Can urban indicators predict home price appreciation? Implications for redlining research', *Real Estate Economics*, Vol. 25(1), pp. 81-104.

Lianos, M. and Douglas, M. (2000), 'Dangerization and the end of deviance: the institutional environment', *British Journal of Criminology*, 40, pp. 261-78.

Liberty (2002), 'Comment by Sadiq Khan', Autumn, p. 3.

Light, J. (2002), 'Urban Security from Warfare to Welfare', *International Journal of Urban and Regional Research*, Vol. 26(3), pp. 607-13.

Linder, R. (1996), *The reportage of urban culture*, Cambridge University Press, Cambridge.

Lindsay, J.M. (1997), *Techniques in Human Geography*, Routledge, London.

Lisle-Williams, M. (1984), 'Beyond the market: the survival of family capitalism in the English merchant banks', *British Journal of Sociology*, XXXV, pp. 241-71.

Llewelyn-Davies (1996), *Four World Cities: a comparative study of London, Paris, New York and Tokyo*, Government Office for London, London.

Lloyd's List (1993), 'Explosion tests bomb pool resolve', 26 April, p. 8.

Loader, I. (1997), 'Thinking normatively about private security', *Journal of Law and Society,* Vol. 34(3), pp. 377-94.

London Chamber of Commerce and Industry (1994), *Invest in London - an international city,* EMP plc, London.

London Chamber of Commerce and Industry (1996), *Invest in London - an international city,* EMP plc, London.

Longcore, T.R and Rees, P.W. (1996), 'Information technology and downtown restructuring: the case of New York City's financial district', *Urban Geography*, Vol. 17(4), pp. 354-72.

Low, V. and Pryer, N. (1993), '£1 billion claims may be double true figure', *Evening Standard*, 30 April, p.6.

Lowenthal, D. (1975), 'Past time, present place: landscape and memory', *Geographical Review*, 65, pp. 1-36.

Lowther, J. (1999), 'Institutional change and its role in strategic economic development: the case of the Tees Valley', paper to a seminar on *Institutional Capacity, Social Capital and Urban Governance*, Newcastle, April.

Luymes, D. (1997), 'The fortification of suburbia: investigating the rise of enclave communities', *Landscape and Urban Planning*, 39, pp. 187-203.
Lyon, D. (1994), *The electronic eye - the rise of surveillance society*, Polity Press, Cambridge.
Lyon, D. (2002a), 'Everyday surveillance: personal data and social classifications', *Information, Communication & Society*, Vol. 5(2), pp. 242-57.
Lyon, D. (2002b), 'Technology vs Terrorism: ID cards, CCTV and biometric surveillance of the city', paper presented at the *Cities as Strategic Sites* conference, University of Salford, November.
Macleod, G. and Goodwin, M. (1999), 'Space, scale and state strategy: rethinking urban and regional governance', *Progress in Human Geography*, Vol. 23(4), pp. 503-527.
Malmberg, T. (1980), *Human territoriality: a survey of the behavioural territories of man with preliminary analysis and discussion of meaning*, Aldine, Chicago.
Madanipour, A. (1995), 'Reading the city', in Healey *et al* (ed.), *Managing cities: the new urban context*, John Wiley, London.
Madanipour, A. (1996), *Design of urban space: an inquiry into a socio-spatial process*, John Wiley, Chichester.
Marcuse, P. (1993), 'What's so new about divided cities', *International Journal of Urban and Regional Research*, 17, pp. 353-365.
Marcuse, P. (1997), 'Walls of fear walls of support', in Ellin, N. (ed.) *Architecture of Fear*, Princeton Architectural Press, New York.
Marcuse, P. (2001), 'Reflections on the Events', City, Vol.5(3), pp. 394-7.
Marcuse, P. (2002a), 'Afterword', in Marcuse, P. and van Kempton, R., *Of States and Cities: The Partitioning of Urban Space*, Oxford University Press, Oxford, pp. 269-82.
Marcuse, P. (2002b), 'Urban Form and Globalization after September 11th: The View From New York, *International Journal of Urban and Regional Research*, Vol. 26(3), pp. 596-606.
Marcuse, P. and van Kempton, R. (2002), *Of States and Cities: The Partitioning of Urban Space*, Oxford University Press, Oxford.
Marston, S.A. (1997), 'Who's policing what space? Critical silences in Steve Herbert's Policing Space', *Urban Geography*, Vol. 18(5), pp. 385-8.
Marx, G.T. (1985), 'The surveillance society - the threat of 1984-style techniques', *The Futurist*, June, pp. 21-26.
Mayhew, J. (2001), 'Address on Management', *Presentation at Cambridge University*, 15 November 2001.
Mayhew, P. (1979), 'Defensible space: the current status of crime prevention theory', *The Howard Journal*, 18, pp. 150-59.
McCahill, M. and Norris, C. (2002), *CCTV in Britain*, Working Paper 3, Centre for Criminology and Criminal Justice, University of Hull.
McDowell, L. (1994), 'Social justice, organizational culture and workplace democracy: cultural imperialism in the City of London', *Urban Geography*, Vol. 15(7), pp. 661-80.
McDowell, L. (1999), 'City life and difference: negotiating diversity', in Allen, J., Massey, D. and Pryke, M. (eds.) *Unsettling Cities*, Routledge/Open University, London.
McKane, L. (1975), *Security restrictions in Omagh: an analysis of the direct effects on traffic and indirect effects on the commercial and social functions of Omagh*, unpublished undergraduate dissertation, Geography Department, Queens University, Belfast.
McKenzie, R. (1968), *On Human Ecology*, University of Chicago Press, Chicago.
McLaughin, E. and Muncie, J. (1999), 'Walled cities: surveillance, regulation and segregation', in Pile, S., Brook, C. and Mooney, G. (eds.), *Unruly Cities?*, Routledge/Open University Press, London.
McNeill, D. (2002), 'Livingstone's London: Left Politics and the World City', Regional Studies, Vol. 36(1), pp. 75-91.

McQuillen, R. (1975), *Ballymena town centre 1974: an example of an premature pedestrian precinc*, unpublished undergraduate dissertation, Geography Department, Queen University, Belfast.

Menon, J (1996), 'Labour slates postcodes', *Insurance Age,* December, p. 9.

Merari, A. (1993), 'Terrorism as a strategy of insurgency', *Terrorism and Political Violence,* Vol 5(4), Winter, pp. 213-51.

Mercer, C. (1975), *Living in cities: psychology and the urban environment*, Penguin, Harmondsworth.

Merrifield, A. (1997), 'Public space: integration and exclusion in urban life', *City*, 5-6, pp. 57-72.

Merrifield, A. and Swyngedouw, E. (eds.) (1997), *The Urbanization of Injustice*, Lawrence and Wishart, London.

Merrifield, A. and Swyngedouw, E. (1997), 'Social justice and the urban experience'. In Merrifield and Swyngedouw (eds.), *The Urbanization of Injustice*, Lawrence and Wishart, London.

Milgram, S. (1970), 'The experience of living in cities', *Science*, 167, pp. 1461-8.

Miller, J. and Glassner, B. (1997), 'The inside and outside - finding realities in interviews, in Silveramn, D. (ed.), *Qualitative Research*, Sage, London.

Mills, E.S. (2002), 'Terrorism and U.S. Real Estate', *Journal of Urban Economics*, 51, pp. 198-204.

Millward, D. (1996), 'TV cameras have limited effect on town centre crime', *Electronic Telegraph,* 2 January.

Mishler, E.G. (1991), *Research interviewing - context and narrative*, Harvard Press.

Mitchell, D. (1997), 'Power, tactics and the political geography of policing: Comments on Steve Herbert's policing space', *Urban Geography*, Vol. 18(5), pp. 392-7.

Mitchell, W.J.T. (ed.) (1994), *Landscape and Power*, University of Chicago Press, Chicago.

Mollenkopf, J.E. (1993), *Urban nodes in the global system*, Social Science Research Council, New York.

Moore, R. (1996), 'Record-breaking tower would lift City', *Electronic Telegraph*, Issue 475, 10 September.

Morimiya (1985), 'Covering natural disasters in the Japanese market', *Risk Management*, Vol. 32(12), pp. 18-26.

Morris, A.E.J (1994), *History of urban form - before the industrial revolution* (Third Edition), Longman, London.

Morris, D. (1967), *The naked ape*, McGraw-Hill, New York.

Mumford, L. (1961), *The City in History*, Penguin, Harmondsworth.

Murray, I. (1994), 'Insurers blacklist Inner-city estates', *The Times*, 25 March.

Murray, R. (1982), 'Political violence in Northern Ireland 1969-1977', in Boal, F.W. and Douglas, J.N.H. (eds.). *Integration and division: geographical perspectives on the Northern Ireland problem*, Academic Press, London, pp. 309-31.

Myers, K. (1996), 'Irelands ceasefire is over', *The Times*, 3 January, p. 16.

Nacos, B.L. (2002), *Mass-mediated terrorism: The central role of the media in terrorism and counter-terrorism*, Rowman and Littlefield, Lanham.

Napier, M. (2000), 'The state of human settlements in South Africa: the impact of five years of democratic rule', University of Newcastle, *School of Architecture, Planning and Landscape Research Seminar Series*, 4 May.

Nash, C. (1999), 'Landscapes', in Cloke, P., Crang, P. and Goodwin, M. (eds.), *Introducing Human Geographies*, Arnold, London, pp. 217-25.

Neill, W. (1992), 'Re-imaging Belfast', *The Planner*, 2 October, pp. 8-10.

Neill, W.J.V., Fitzsimons, D.S. and Murtagh, B. (1995), *Reimaging the Pariah City - Urban development in Belfast and Detroit*, Avebury, Aldershot.

Neill, W.J.V. (1995), 'Lipstick on the gorilla? Conflict management, urban development and image making in Belfast', in Neill, W.J.V., Fitzsimons, D.S. and Murtagh, B., *Reimaging the Pariah City - Urban development in Belfast and Detroit*, Avebury, Aldershot, pp. 50-76.

Newman, O. (1972), *Defensible space - crime prevention through urban design*, Macmillan, New York.

Newman, O. (1973), *Defensible space: people and design in the violent city*, Architectural Press, London.

Newman, O. (1980), *Community of interest*, Anchor Press, New York.

Newman, O. (1995), 'Defensible space - a new physical planning tool for urban revitalisation', *Journal of the American Planning Association*, Vol. 61(2), Spring, pp. 149-155.

Newman, O. (1996), *Creating defensible space*, US Department of Housing and Urban Development - Office of Policy Development and Research.

Newman, O. (1997), 'Defensible space', *National Housing Institute*, Shelterforce Online, May/June.

Newman, P. and Thornley, A. (1997), 'Fragmentation and centralisation in the governance of London: influencing the urban policy and planning agenda', *Urban Studies*, Vol. 34(7), pp. 967-88.

Norfolk, S (1994), 'Houston streets: a world apart', *The Independent*, 9 November, p. 26.

Norman, P. (1975), Managerialism: a review of recent work', in Harloe (ed.) *Proceedings of the conference of urban change and conflict*, pp. 62-98.

Norris, C. and Armstrong, G. (1999), *The maximum surveillance society: The rise of CCTV*, Berg, Oxford.

Notton, J. (1997), 'The use of technology to combat terrorist crime in the City of London', *City Security*, 13, Spring, pp. 15-16.

O'Brien, B. (1995), *The Long War*, The O'Brien Press, Dublin.

O'Brien, B. (1997), *A Pocket History of the IRA*, The O'Brien Press, Dublin.

O'Dowd, L. (1998), 'Coercion, territoriality and the prospects for a negotiated settlement in Ireland', *Political Geography*, Vol. 17(2), pp. 239-49.

Oakley, R.B. (1995), *Recent trends in domestic and international terrorism*, Center for National Security Studies.

Oc, T. and Tiesdell, S. (1997), *Safer city centres - reviving the public realm*, Paul Chapman, London.

Olnis, R. (1992), 'Prince steps in over City bomb cover', *The Sunday Times*, 13 December.

Pahl, R.E. (1977b), 'Managers, technical experts and the state: forms of mediation, manipulation and dominance in urban and regional development', in Harloe, M. (ed.) *Captive Cities*, Wiley, Chichester.

Pahl, R.E. (1969), Urban social theory and research', *Environment and Planning A*, 1, pp. 143-53.

Pahl, R.E. (1970), *Whose City? And other essays on sociology and planning*, Longman, London.

Pahl, R.E. (1975), *Whose City? And further essays on urban sociology (second edition)*, Penguin, Middlesex.

Pahl, R.E. (1977a) 'Collective consumption and the state in capitalist and state socialist societies', in Scase, R. (ed.) *Industrial society: class, cleavage and control*, Allen and Unwin, London.

Pain, R.H. (1995), Local contexts and fear of crime: elderly people in North East England, *Northern Economic Review*, 24, pp. 96-109.

Paletz, D.L. and Schmid, A.P. (ed.) (1992), *Terrorism and the media*, Sage, California.

Park, R.E. (1929), 'The city as a laboratory', in Smith T. and White, L. (eds.), *Chicago: an experiment in social science research*, University of Chicago Press, Chicago, pp. 24-37.
Patel, S. and Hamnet, C. (1987), How insurers strangle cities, *New Society*, 11 December.
Pawley, M. (1998), *Terminal Architecture*, Reaktion, London.
Payne, S. (1999), 'Interview in qualitative research', in Memon, A. and Bull, R. (eds.) (1999), *The handbook of the psychology of interviewing*, Wiley, London, pp. 89-103.
Pendlebury, R. (1993), 'Insurance wipeout', *Daily Mail*, 26 April, p. 9.
Perle, E.D., Lynch, K. and Horner, J. (1994), 'Perspectives on mortgage lending and redlining', *Journal of the American Planning Association*, Vol. 60(3), pp. 344-54.
Phelps, N. and Tewder-Jones (1998), 'Institutional capacity building in a strategic policy vacuum: the case of LG electronics in South Wales', paper presented at the *Reflections on the institutional turn in local economic development conference* at the University of Sheffield, 9-10 September.
Philo, C. and Kearns, G. (1993), *Selling places: The city as cultural capital, past and present*, Pergamon, Oxford.
Picard, R.G. (1994), 'The maturation of communication and terrorism studies', *Journal of Communication*, Vol. 44(1), Winter, pp. 122-127.
Pickavance, C. (1982), 'Review of "Social theory and the urban question" and "City, class, and capital"', *Critical Social Policy*, 2, pp. 94-8.
Pickavance, C. (1984), 'The structuralist critique in urban studies' in Smith, M. (ed.), *Cities in transformation: class capital and the state*, Sage, London.
Pile, S., Brook, C. and Mooney, G. (1999), *Unruly Cities*, Routledge, London.
Police Authority for Northern Ireland (1995), *Opinion poll on CCTV*.
Poole, R. and Williams, D. (1996), 'Success in a surveillance society', *Security Management*, May, pp. 29-35.
Porteous, J.D. (1976), 'Home: the territorial core', *Geography Review*, 66, pp. 383-90.
Postgate, J.N. (1992), *Early Mesopotamia*, Routledge, London.
Powell, W.W. and Dimaggio, P.J. (1991), *The new institutionalism in organisational analysis*, Chicago University Press, Chicago.
Power, D. (2001), 'Technology and the structuring of the financial district of London', in Brunn, S. and Leinbach, S. (eds.) *The Worlds of Electronic Commerce*, Wiley Chichester.
Power, D. (2001), 'Working and voting for a world class city? A critical viewpoint on the Corporation of London's Place in London Governance', *The London Journal*, Vol. 26(2), pp. 51-64.
Power, L. (1993), 'On the road to a better future', *City Post*, 21 April, p. 1.
Poyner, B. (1983), *Design against crime - beyond defensible space*, Butterworths, London.
Pratt, K. (1993), 'The fate of Pool Re', *The CII Journal*, 20-22 September.
Pridham, G. (1987), *Political parties and coalition behaviour in Italy: an interpretive study*, Croom Helm, London.
Pryke, M. (1991), 'An international city going global: spatial change in the City of London', *Environment and Planning D: Society and Space*, Vol. 9, pp. 197-222.
Pryke, M. (1994), 'Looking Back on the space of a boom: (re)developing spatial matrices in the City of London', *Environment and Planning A*, Vol. 26, pp. 235-264.
Pryke, M. and Lee, R. (1995), 'Place your bets: towards an understanding of globalisation, socio-financial engineering and competition within financial centres', *Urban Studies*, Vol. 32(2), pp. 329-44.
Quinn, D. (1993), *Understanding Northern Ireland*, Baseline Books, London.
Raco, M. (1998), 'Assessing "institutional thickness" in the local context: a comparison of Cardiff and Sheffield', *Environment and Planning A*, 30, pp. 975-96.
Raco, M. (1999), 'Researching the new urban governance: an examination of closure, access and complexities of institutional research', *Area*, Vol. 31(3), pp. 271-279.

Raper, J., Rhind, D. and Shepard, J. (1992), *Postcodes: a new geography*, Longman, Harlow.
Rappoport, A. (1977), *Human aspects of urban form*, Pergamon, Oxford.
Rhodes, R.A.W. (1997), *Understanding governance: policy networks, governance, reflexivity and accountability*, Open University Press.
Richardson, T. (1996), 'Foucauldian discourse: power and truth in urban and regional policy making', *European Planning Studies*, Vol. 4(3), pp. 279-92.
Robertson, R. (1992), 'Globality and Modernity', *Theory, Culture and Society*, Vol. 9, pp. 153-61.
Robertson, R. (1992), *Globalization*, London, Sage.
Robins, K. (1991), 'Prisoners of the city: whatever could a postmodern city be?', *New Formations*, 15, Winter, pp. 1-23.
Rogers, P. (1996), *Economic Targeting and Provisional IRA Strategy*, University of Bradford, Department of Peace Studies, Paper 96.1.
Ronchek, D.W. (1981), 'Dangerous places: crime and the residential environment', *Social Forces*, 60, pp. 74-96.
Rose, N. (1996), 'The death of the social? Refiguring the territoriality of government', *Economy and Society*, 25, pp. 327-56.
Rosen, B. (2001), 'A cautionary tale for a new age of surveillance', *New York Times*, 7 October.
Rotenburg, R. (1993), 'Introduction', in Rotenburg, R. and McDonagh, G. (eds.), *The Cultural Meaning of Urban Space*, Bergin and Garvey, London.
Rowley, G. (1992), 'Human space, territoriality, and conflict: an exploratory study with special reference to Israel and the West Bank', *Canadian Geographer*, 36, pp. 210-21.
Rubenstein, J. (1973), *City police*, Ballentine, New York.
Ryan, M. (1994), *War and Peace in Ireland - Britain and the IRA in the New World Order*, Pluto-Press, London.
Rycus, M.J. (1991), 'Urban terrorism: A comparative study', *The Journal of Architectural and Planning Research*, Vol. 8(1), Spring, pp. 1-10.
Sack, R.P. (1986), *Human Territoriality - its theory and history*, Cambridge University Press, Cambridge.
Sack, R.P. (1992), *Place, modernity, and the consumer's world - a relational framework for geographical analysis*, Johns Hopkins University Press, Baltimore?.
Saifer, M. (2001), 'Confronting "Urbicide", *City*, Vol. 5(3), pp. 416-29.
Sassen, S, (1991), *The global city*, Princetown University Press, Princetown.
Sassen, S. (1993), *Cities in the world economy*, Sage, London.
Sassen, S. (2000), 'New frontiers facing urban sociology at the Millennium', *British Journal of Sociology*, Vol. 15(1), pp. 143-59.
Saunders, P. (1979), *Urban politics: a sociological interpretation*, Hutchinson, London.
Saunders, P. (1981), *Social theory and the urban question (second edition)*, Unwin Hyman, London.
Savage, M. and Warde, A. (1993), *Urban sociology, capitalism and modernity*, Macmillan, London.
Savitch, H.V. and Ardashev, G. (2001), 'Does terror have an urban future', *Urban Studies*, Vol. 38(13), pp. 2515-33.
Schachter, R. (1981), *Insurance redlining: organizing to win*, National Training and Information Center, Chicago.
Schaffert, R.W. (1992), *Media coverage and political terrorists: A quantitative analysis*, Praeger, New York.
Schmalz, J. (1988), 'Fearful and angry Floridians erect street barriers to crime', *New York Times*, 6 December, City Edition, p. A-1.

Schmid, A.P. (1992), 'The response problem as a definitional problem', in Schmid, A.P. and Crelinsten, R.D. (ed.) *Terrorism and Political Violence, Special Edition on Western Responses to Terrorism*, Frank Cass, London.

Scott, A.J. and Stroper, M. (1986), *Production, work and territory: the geographical anatomy of industrial capitalism*, Allen and Unwin, London.

Scott, A.J. and Soja, E.W. (eds.) (1997), *The city of Los Angeles and urban theory at the end of the twentieth century*, University of California Press, Berkeley.

Scott, M. (1996), 'Hope yet for policy pariahs', *The Observer*, 6 November.

Seley, J.E. and Wolpert, J. (1982), 'Negotiating urban risk', in Herbert, D.T. and Johnson, R.J. (eds.), *Geography and the urban environment - progress in research and applications*, Volume V, John Wiley, London, p. 279.

Sharrock, D. (1996a), 'Threat of new loyalist terror', *The Guardian*, 13 March, p. 1.

Sharrock, D. (1996b), 'Armed loyalist gang seizes £1m in Belfast', *The Guardian*, 15 April, p. 3.

Sharrock, D. (1996c), 'Loyalist bomb hoax in Dublin', *The Guardian*, 7 May, p. 4.

Sheppard, A.W. (1976), *Business as usual? - a geographical analysis of shopping behaviour in Belfast city centre (1969-75)*, unpublished undergraduate dissertation, Queens University, Belfast.

Shirlow, P. (1998), *Fundamentalist Loyalism: discourse, dialogue and defence*, paper presented at the *Landscapes of Defence conference*, Oxford Brookes University, 31 May.

Short, J.R. (1996), *The Urban Order*, Blackwell, Oxford.

Sibley, D. (1990), 'Invisible women? The contribution of the Chicago school of social service administration to urban analysis', in Anderson, K. and Gale, F. (eds.) *Inventing places: studies in cultural geography*, Longman, Cheshire, pp. 107-22.

Sibley, D. (1995), *Geographies of Exclusion: society and difference in the West*, Routledge, London.

Sivell, G. (1993), 'Walled City mooted to thwart the terrorists', *The Times*, 27 April, p. 23.

Sjoberg, G. (1960), *The pre-industrial city: past and present*, Free Press, Glencoe.

Smart, B. (1993), *Postmodernity*, Routledge, London.

Smith, G. (1994), 'London's great security ring stays in spite of ceasefire', *Evening Standard*, 1 September.

Smith, H. (1993), 'Checkpoints mark fortress London', *Evening Standard*, 7 June, p. 6.

Smith, K. (1992), *Environmental hazards - assessing risk and reducing disaster*, Routledge, London.

Smith, M. (1992), 'State to back insurers over bomb damage', *Daily Telegraph*, 21 December, p. 2.

Smith, N. (1996), *The new urban frontier: gentrification and the revanchist city*, Routledge, London.

Soffer, A. and Minghi, J.V. (1986), 'Israel's security landscapes: the impact of military considerations on the land uses', *Professional Geographer*, 38, pp. 28-41.

Soja, E. (1997), 'Margin Alia: Social Justice and the New Cultural Politics'. In Merrifield and Swyngedouw (eds.), *The Urbanization of Injustice*, Lawrence and Wishart, London.

Soja, E.W. and Scott, A.J. (1986), 'Los Angeles: Capital of the late twentieth century', *Environment and Planning D: Society and Space*, 4, pp. 249-54.

Soja, E.W. (1980), 'The sociospatial dialectic', *Annals of the Association of American Geographers*, 70, pp. 207-25.

Soja, E.W. (1989), *Postmodern Geographies - The reassertion of space in critical social theory*, Verso, London.

Soja, E.W. (2000), *Postmetropolis*, Blackwell, Oxford.

Sorkin, M. (ed.) (1995), *Variations on a theme park - the new American city and the end of public space*, Hill and Wang, New York.

Squires, G.D., DeWolfe, R. and DeWolfe, A.S. (1979), 'Urban decline or disinvestment: uneven development, redlining and the role of the insurance industry', *Social Problems*, Vol. 27(1), October, pp. 79-95.
Squires, G.D. and Valez, W. (1987), 'Insurance redlining and the transformation of the urban metropolis', *Urban Affairs Quarterly*, Vol.23(1), September, pp. 63-83.
Squires, G.D. and Valez, W. (1988), 'Insurance redlining and the process of discrimination', *Review of the Black Political Economy*, 16, Winter, pp. 63-75.
Squires, G.D., Valez, W. and Taeuber, K.E. (1991), 'Insurance redlining, agency location, and the process of urban disinvestment', *Urban Affairs Quarterly*, Vol. 26(4), pp. 67-88.
Stewart, J. (1998), 'Peddling fear', *New Times Los Angeles*, 19 November.
Stoker, G. (1996), 'Introduction, normative theory of local government and democracy', in King, D. and Stoker, G. (eds.), *Rethinking local democracy*, Macmillan, London, pp. 1-27.
Sudjic, D. (2001a), 'A thoroughly modernising major', *The Observer (online)*, 8 July.
Sudjic, D. (2001b), 'Save Spitalfield from market forces', *The Observer (online)*, 15 July.
Sudman, S. and Kalton, G. (1986), 'New developments in the sampling of special populations', *Annual Review of Sociology*, 12, pp. 401-29.
Suh, S.T. (1998), *Evaluating the capacity of plan making in Seoul through institutional analysis*, unpublished PhD thesis, Department of Town and Country Planning, University of Newcastle.
Suisman, D.R. (1989), *Los Angeles Boulevard*, Los Angels Forum of Architecture and Urban Design.
Sutcliffe, M. (1996), 'The fragmented city: Durban, South Africa', *International Journal of Social Sciences*, Vol. 48(1), pp. 67-72.
Swanson, G.E. (1992), 'Modernity and the postmodern', *Theory, Culture and Society*, Vol. 9, pp. 147-51.
Swanstrom, T. (2002), 'Are fear and urbanism at war?', *Urban Affairs Review*, Vol. 38(1), pp. 135-40.
Swyngedouw, E. (1997), Neither global or local: 'glocalisation' and the politics of scale, in Cox, K (ed.) *Spaces of Globalization*, Guilford Press, New York, pp. 137-66.
Taylor, R. (1988), 'Social scientific research on the "troubles" in Northern Ireland', *Economic and Social Review*, 19, pp. 123-45.
Taylor, R.B. (1988), *Human territorial functioning*, Cambridge University Press, Cambridge.
Thielgaard-Watts, M. (1957), *Reading the landscape - an adventure in ecology*, Macmillan, New York.
Thompson, R. and Waters, R. (1993), 'Japanese institutions to review security', *Financial Times*, 27 April.
Thomson, A. (1996), 'Sinn Fein told to end spate of killings', *The Times*, 8 January 1996, p. 2.
Thornley A. (1998), 'Institutional change and London's urban policy agenda', *Annals of Regional Science*, Vol. 32(1), pp. 163-83.
Threadgold, S. (1995), 'Spot the difference', *Post Magazine*, 29 June.
Threadgold, S. (1996), 'Inner city pressure', *Post Magazine*, 8 February.
Thrift, N. (1994), 'On the social and cultural determinants of international financial centres: the case of the City of London', in Corbridge, S, Martin R and Thrift, N. (eds.), *Money, Space, Power*, Blackwell, Oxford, pp. 327-355.
Tijerino, R. (1998), 'Civil spaces: a critical perspective of defensible space', *Journal of Architectural and Planning Research*, 15, pp. 321-37.
Todtling, F (1995), 'The uneven landscape of innovation poles: local embeddedness and global networks', in Amin, A. and Thift, N. (eds.), *Globalization, Institutions and Regional Development in* Europe, Oxford University Press, Oxford, pp. 68-90.

Tomaney, J. (2001), 'The new governance of London: a case of post democracy?', City, Vol. 5, pp. 225-48.

Tootell, G.M.B. (1996), 'Redlining in Boston - do mortgage lenders discriminate against neighbourhoods?', *Quarterly Journal of Economics*, Vol. 111(4), pp. 1049-79.

Townsend, M. and Harris, P. (2003), 'Security role for traffic cameras', *The Observer*, 9 February.

Trancik, R. (1986), *Finding lost space: theories of urban design*, Van Noostrand Reinhold, New York.

Tranter, P. and Parkes, D. (1979), 'Time and images in urban space', *Area*, 11, pp. 115-20.

Travers, T. and Jones, G. (1997), *The new government of London*, Joseph Rowntree Foundation, York.

Travers, T., Jones, G., Hebbert, M. and Burnham, J. (1991), *The Government of London*, Joseph Rowntree Foundation, York.

Treanor, J. (2001), 'Insurers work with government on new terms for terrorism claims', *Electronic Telegraph*, 22 December.

Tuan, Y.F (1979), *Landscapes of Fear*, Pantheon Books, New York.

Tuan, Y.F. (1979), 'Thought and Landscape', in Meinig, D.W. (ed.) (1979), *The Interpretation of Ordinary Landscapes*, Oxford University Press, Oxford.

Urban Task Force (1999), *Towards an Urban Renaissance*, E. & F.N. Spon, London.

Urry, J. (2002), 'The Global Complexities of September 11th', Theory, Culture and Society, Vol. 19(4), pp. 57-69.

USA Today (2002), 'Gated Communities more popular and not just for the rich', 15 December.

Valler, P. (1999), 'They say it's Europe's richest inner city. Try telling them that in Hackney', *Independent (online)*, 11 February.

van Aartijk Jr, P. (2001), 'Can terrorism insurance pool calm insurers fears?', *Independent Agent*, December.

Vidler, A. (2001a), 'The City Transformed: Designing Defensible Space', *New York Times*, 23 September.

Vidler, A. (2001b), Lecture to the American Collegiate Schools of Architecture Admissions Conference, 7 November, New York.

Wadham, J. (1996), Letter to *The Times*, 4 April.

Wagner, A.E. (1997), 'A study of traffic pattern modification in an urban crime prevention program', *Journal of Criminal Justice*, Vol. 25(1), 19-30.

Wagner, P.L. and Mikesell, M.W. (eds.) (1962), *Readings in Cultural Geography*, University of Chicago Press, Chicago.

Warren, R. (2002), 'Situating the City and September 11th: Military Urban Doctrine, "Pop-up" Armies and Spatial Chess', *International Journal of Urban and Regional Research,* Vol. 26(3), pp. 614-9.

Watson, S. and Gibson, K. (eds.) (1994), *Postmodern cities and spaces*, Routledge, London.

Watt, N. (1995), 'Former IRA man shot dead for drug dealing', *The Times*, 20 December, p. 4.

Watt, N. and Hicks, J. (1996), 'Peace in jeopardy as Mayhew links IRA to killings', *The Times*, 3 January, p. 2.

Webb, G. (1996), 'How bombers use public transport to beat checkpoints', *Evening Standard*, 19 February.

Weber, M. (1947), *The theory of social and economic organisation*, Free Press, Glencoe, Illinois.

Wekerle, G.R. and Whitzman, C. (1995), *Safe cities - guidelines for planning, design, and management*, International Thompson Publishing Europe, London.

Welsh, F. (1986), *Uneasy City - an insiders view of the City of London*, Weidenfield and Nicolson, London.

Wenban-Smith, A. (1999), Sustainable institutional capacity for planning: summary note', paper to a seminar on *Institutional capacity, social capital and urban governance*, Newcastle, April.
Werlen, B. (1993), *Society, action and space: an alternative human geography*, Routledge, London.
White, G.F. (1942), *Human adjustment to floods: a geographical approach to flood problems in the United States*, Research Paper 29, Department of Geography, University of Chicago.
White, J.R. (1991), *Terrorism – an introduction*, Brooks/Cole, California.
Widgery, D. (1991), *Some lives! A GP's East End*, London.
Wilkinson, P. (1996), 'Blood is spilling over the map', *The Observer*, 21 July.
Wilkinson, P. (1997), 'The media and terrorism: a reassessment', *Terrorism and Political Violence*, Vol. 9(2), pp. 51-64.
Williams, K.S. and Johnston, C. (2000), 'The politics of selective gaze: closed circuit television and the politics of public space', *Crime, Law and Social Change*, 34, pp. 183-210.
Williams, P. (1978), 'Urban managerialism: a concept of relevance', *Area*, 10, pp. 236-40.
Williams, P. (1982), 'Restructuring urban managerialism: Towards a political economy of urban allocation', *Environment and Planning A*, Vol. 14(1), pp. 95-105.
Williams, P. and Dickinson, J. (1993), 'Fear of crime: read all about it', *British Journal of Criminology*, Vol. 33(1), pp. 33-56.
Williams, S. (1991), 'The coming of the groundscapers', in Budd, L., and Whimster, S. (eds.) *Global Finance and Urban Living: a study of metropolitan change*, Routledge, London, pp. 246-59.
Williams, T. (2001), 'We cannot surrender the city', *Regeneration and Renewal*, 21 September, p. 14.
Wilson, D. (1989), 'Towards a revised urban managerialism: local managers and community development block grants', *Political Geography Quarterly*, Vol. 8(3), pp. 21-41.
Wilson, D. (1993), 'Organisational perspectives and urban spatial structure: a review an appraisal', *Journal of Planning Literature*, Vol. 7(3), pp. 227-37.
Wirth, L. (1938), 'Urbanism as a way of life', *American Journal of Sociology*, 44, pp. 1-24.
Wittfogal, K.A. (1957), *Oriental despotism: a comparative study of total power*, Yale University Press, New Haven.
Wolch, J. and Dear, M. (1989), *The power of geography - how territory shapes social life*, Unwin Hyman, London.
Wong, C. (1997), 'Crime risk in urban neighbourhood: the use of insurance data to analyse changing spatial forms', *Area*, Vol. 29(3), pp. 228-40.
Wood, D. (2002), 'Militarisation, modernisation or myth? Algorithmic surveillance technologies and the automatic production of urban space', paper presented at the *Cities as Strategic Sites* conference, University of Salford, November.
Woods, M. (1996), 'Spatialising elite theory: a cultural geography of local politics', paper presented at the *RGS-IBG annual conference*, University of Strathclyde, 6 January.
Worden, N. (1994), *The making of modern South Africa: conquest, segregation and apartheid*, Blackwell, Oxford.
Wright, J.K. (1947) '*Terrae incognitae*: the place of the imagination in geography', *Annals of the Association of American Geographers*, 37, pp. 1-15.
Yeoh, B.S.A. (1999), 'Global/globalizing cities', *Progress in Human Geography*, Vol. 23(4), pp. 607-16.
Zukin, S. (1982), *Loft living: culture and capital in urban change*, Johns Hopkins University Press, Baltimore.

Zukin, S. (1988), 'The postmodern debate over urban form', *Theory, Culture and Society*, Vol. 5(2-3), June, pp. 431-46.

Zukin, S. (1992), 'The city as a landscape of power; London and New York as global financial capitals', in Budd, L. and Whimster, S. (eds.), *Global finance and urban living - a study of metropolitan life*, Routledge, London.

Zukin, S. (1995), *The culture of cities*, Blackwell, Oxford.

Zulaika, J. and Douglass, W.A. (1996), *Terror and Taboo: The follies, fables and faces of terrorism*, Routledge, New York.

Zwingle, E. (1991), 'Docklands - London's new frontier', *National Geographic*, July, pp. 32-59.

Index